2

Band 20 · Februar 1984

isotopenpraxis

AKADEMIE-VERLAG

ISSN 0021–1915 Isotopenpraxis, Berlin **20** (1984) 2, 41–80

EVP 9,– M

An unsere Autoren

Die Zeitschrift veröffentlicht Beiträge zu folgenden Themen: Kontrolle und Analyse radioaktiver und stabiler Isotope, Anwendung radioaktiver und stabiler Isotope in allen Gebieten der Forschung und Technik, der Industrie, Landwirtschaft und Medizin, Herstellung radioaktiver und stabiler Isotope sowie von markierten Verbindungen, kernelektronische Geräte für die Isotopenanwendung (Dickenmesser, Dichtemesser u. ä.) Detektoren für isotopentechnische Arbeiten und den Strahlenschutz in Isotopenlaboratorien, Meßtechnik und Strahlenschutz, Dosimetrie in Isotopenlaboratorien, kerntechnische Geräte für die Isotopenanwendung (Beton, Bestrahlungsanlagen u. ä.), Strahlenchemie, Radiochemie, ökonomische und perspektivische Fragen, Fragen der Weiterbildung.

1. Aufnahme von Manuskripten

Zur Veröffentlichung angenommen werden nur druckreife, noch nicht anderweitig publizierte Beiträge in deutscher, englischer oder russischer Sprache zum Themenbereich der Zeitschrift und zwar

- Originalarbeiten
- Kurze Mitteilungen
- Aus der Praxis
- Zusammenfassende Berichte
- Tagungsberichte
- Informationen.

Sind der Beitrag oder Teile daraus bereits veröffentlicht oder ist eine anderweitige Veröffentlichung vorgesehen, so muß dies der Redaktion mit genauer Angabe der Zweitveröffentlichung mitgeteilt werden. Mit der Annahme des Manuskriptes behält sich der Verlag das Recht der Vervielfältigung, Verbreitung und Übersetzung innerhalb der gesetzlichen Schutzfrist vor. Auszugsweiser Nachdruck oder photomechanische u. a. Wiedergabe sind an eine schriftliche Genehmigung des Verlages gebunden. Manuskripte, Zeichnungen und Photos von veröffentlichten Arbeiten werden nur auf den besonderen und bis zum Zeitpunkt der Veröffentlichung mitgeteilten Wunsch des Autors zurückgesandt.

Bei der unverlangten Einreichung von Manuskripten wird empfohlen, eine Befürwortung eines der Herausgeber oder Redaktionsausschußmitglieder beizufügen.

Manuskripte sind einzureichen bei der Redaktion, 1086 Berlin, Leipziger Str. 3–4.

2. Form der Manuskripte

Die Autoren werden gebeten:

2 Manuskripte, eines davon in Original-Maschinenschrift, einseitig und mit 2zeiligem Abstand einzureichen; gut leserliche Photokopien sind zulässig; auf dem 1. Blatt (unter Weglassung der Bezeichnung)

- den Titel der Arbeit
- Vorname-Initialen und Namen sämtlicher Autoren
- gegebenenfalls die Institution, in der sie tätig sind
- eine kurze Zusammenfassung der Arbeit von höchstens 15 Zeilen
- Keywords in alphabetischer Reihenfolge und
- als Fußnote die Anschrift der Institution anzugeben;

wenn möglich, auf gesonderten Blättern Namen, Titel der Arbeit und Zusammenfassung in den beiden anderen obengenannten Sprachen 2fach beizufügen;

Literaturhinweise (fortlaufend numeriert), Fußnoten, Tabellen, Abbildungen (in 2–3facher Veröffentlichungsgröße), Abbildungsunterschriften (getrennt von den Abbildungen mit Zeichenerklärungen, Quellenangaben, falls erforderlich, und gegebenenfalls mit einem Vermerk für besondere Wünsche bezüglich der Abbildungsgröße), private und Dienstanschrift der Verfasser mit akademischen Titeln und ausgeschriebenen Vornamen auf besonderen Blättern ans Ende des Manuskriptes zu stellen;

Wünsche für den Kolumnentitel (Seitenüberschrift) bekanntzugeben;

Absätze durch Einzüge von mindestens 3 Anschlägen kenntlich zu machen; die Maßeinheiten des SI-Systems („Systeme International“) zu verwenden; Korrekturen im Manuskript unmittelbar an der zu korrigierenden Stelle anzubringen;

Formeln deutlich lesbar und möglichst groß einzutragen;

Farbabbildungen und nachträgliche Änderungen oder Ergänzungen in Abbildungen zu vermeiden;

hervorzuhebende Textstellen zu unterstreichen, und zwar halbfetten Satz mit einer geraden schwarzen Linie, kursiven Satz (Schrägschrift) mit einer blauen Wellenlinie (alle Eigennamen). Sperrungen gestrichelt;

besondere Buchstaben durch farbige Unterstreichung zu kennzeichnen, und zwar Formelzeichen in

- Griechisch-Kursiv (α, β, ...) rot,
- Kursiv (A, B, ...) blau
- halbfetter Kursiv ($\boldsymbol{A}$, $\boldsymbol{B}$, ...) 2 Linien blau (Vektoren),
- halbfetter Grotesk (**A**, **B**, ...) 2 Linien violett (Tensoren);

bei Literaturangaben auf Vollständigkeit zu achten. In nachstehender Reihenfolge müssen angegeben sein:

a) bei *Zeitschriften*: Vorname-Initialen und Namen sämtlicher Autoren, Titel der Zeitschrift (nach „Periodica Chimica“ oder möglichst wenig gekürzt), Bandnummer (unterstrichen), Jahreszahl (in Klammern), Heftnummer (nur bei Zeitschriften, die jedes Heft mit S. 1 beginnen) sowie die Seitenzahl, auf der die zitierte Arbeit beginnt.

b) bei *Büchern*: Initialen der Vornamen und Name des Autors (Herausgeber werden durch den Zusatz „ed.“ gekennzeichnet), voll ausgeschriebener Titel des Buches in der Originalsprache, evtl. Bandnummer, voll ausgeschriebener Name des Verlages, Nummer der Auflage (außer der ersten), Erscheinungsort(e), Erscheinungsjahr (hier nicht in Klammern setzen!) und evtl. Kapitelnummer, Seitenzahl u. ä.

c) bei *Sammelbänden, Tagungsberichten*: Es soll zusätzlich der Titel des zitierten Beitrags angegeben werden. Die Seitenzahl muß unbedingt angegeben werden. Aus dem Zitat muß hervorgehen, ob es sich um einen Tagungssammelband oder um einen noch nicht veröffentlichten Vortrag handelt. Zwischen Tagungsort und Erscheinungsort des entsprechenden Tagungsberichts ist zu unterscheiden.

3. Korrekturen

Der Autor erhält zwei Umbruchabzüge, von denen einer möglichst schnell korrigiert an die Redaktion zurückgereicht werden soll. Auf Wunsch des Autors kann die Korrektur auch von der Redaktion gelesen werden.

Durch nachträgliche Manuskriptänderungen verursachte Korrekturkosten gehen, sofern sie 10% der Satzkosten übersteigen, zu Lasten des Autors.

Der Autor erhält 50 Sonderdrucke kostenlos. Weitere Sonderdrucke können bis zur Rückgabe der Korrekturabzüge auf Kosten des Autors bei der Redaktion bestellt werden.

Bezugsmöglichkeiten

Bestellungen sind zu richten

- **in der DDR** an den Postzeitungsvertrieb unter Angabe der Kundennummer des Bestellenden oder an den AKADEMIE-VERLAG, DDR-1086 Berlin, Leipziger Straße 3–4, PF-Nr. 1233;
- **im sozialistischen Ausland** an eine Buchhandlung für fremdsprachige Literatur oder an den zuständigen Postzeitungsvertrieb;
- **in der BRD und Berlin(West)** an eine Buchhandlung oder an die Auslieferungsstelle KUNST UND WISSEN, Erich Bieber OHG, Wilhelmstraße 4–6, D-7000 Stuttgart 1;
- **in den übrigen westeuropäischen Ländern** an eine Buchhandlung oder an die Auslieferungsstelle KUNST UND WISSEN, Erich Bieber GmbH, Dufourstraße 51, CH-8008 Zürich;
- **im übrigen Ausland** an den Internationalen Buch- und Zeitschriftenhandel; den Buchexport, Volkseigener Außenhandelsbetrieb der Deutschen Demokratischen Republik, DDR-7010 Leipzig, Postfach 160; oder an den AKADEMIE-VERLAG, DDR-1086 Berlin, Leipziger Straße 3–4.

Zeitschrift „Isotopenpraxis“

Herausgeber: Im Auftrag des Akademie-Verlages herausgegeben von Prof. Dr. Justus Mühlenpfordt
Verlag: Akademie-Verlag, DDR-1086 Berlin, Leipziger Straße 3–4, PF-Nr. 1233; Fernruf: 2236229; Telex-Nr.: 114420;
Bank: Staatsbank der DDR, Berlin, Kto.-Nr.: 6836-26-20712
Verantwortlich für den Inhalt des Heftes: Prof. Dr. H. Hübner, Leipzig
Anschrift der Redaktion (Fernruf Frau H. Schier: 6351177) der Zeitschrift Isotopenpraxis: Akademie-Verlag, DDR-1086 Berlin, Leipziger Str. 3–4, PF-Nr. 1233
Veröffentlicht unter der Lizenznummer 1527 des Presseamtes beim Vorsitzenden des Ministerrates der Deutschen Demokratischen Republik
Gesamtherstellung: VEB Druckerei „Thomas Müntzer“, DDR-5820 Bad Langensalza
Erscheinungsweise: Die Zeitschrift „Isotopenpraxis“ erscheint monatlich. Die 12 Hefte eines Jahrganges bilden einen Band. Bezugspreis je Band 216,– M zuzüglich Versandspesen (Preis für die DDR 108,– M). Preis je Heft 18,– (Preis für die DDR 9,– M)
Bestellnummer dieses Heftes: 1079/20/2

AN (EDV) 12831

Isotopenpraxis 20 (1984) 2, S. 41 bis 47

Nuclear Analytical Methods in the Semiconductor Industry*)

H. Koch

(Central Institute of Isotope and Radiation Research Leipzig)[1])

The most important radio-analytical methods, used in microelectronics, such as activation analysis, tracer-techniques, autoradiography etc. and their performance limits (dection limit and senitivity) are presented. In this context the detection of areal substance distributions, the determination of depth distributions are discussed other methods of general trace analytics are includet in these considerations for comparison.
Possible applications and the necessary methodological developments and basic researches are derived from the requirements placed by microelectronics on isotope techniques and from the present state of the art. Some results obtained in cooperation with semiconductor-industry are discussed.

Es werden die wichtigsten radio-analytischen Methoden, Aktivierungsanalyse, Tracertechnik, Autoradiography etc., die bei der Entwicklung und Herstellung von Halbleiterbauelementen angewandt werden, beschrieben und deren Nachweisgrenzen und -empfindlichkeiten diskutiert. Von besonderem Interesse sind dabei Probleme der lateralen und Tiefenverteilung von Spurenverunreinigungen und Dotanden. Anhand zahlreicher Beispiele werden neue methodische Entwicklungen sowie deren praktische Anwendung bei der Halbleiterherstellung vorgestellt.

Keywords

autoradiography, depth distribution, etching, integrated circuits, isotope dilution analysis, neutron activation analysis, sensitivity, semiconductor materials, trace amounts, wafer

The production of semiconductor materials, and their application in industry requires sensitive, specific and reproducible analytical techniques. It has been found out that the properties of these materials are very dependent on the presense of impurity or doping atoms in concentrations as small as nanograms per gram (ppb) or smaller.

Fig. 1 shows a part of a 2 inch diameter silison wafer. Such a wafer contains about 250 circuits in its surface. Each of the circuits contains ~70 different devices. A defect at any one of these devices can ruin the whole circuit. For a 100% yield the ~18000 sites must be within specified limits of impurity. Nowadays, slice diameters have alredy reached 4 inches, and a million or more divices on a wafer is not uncommon.

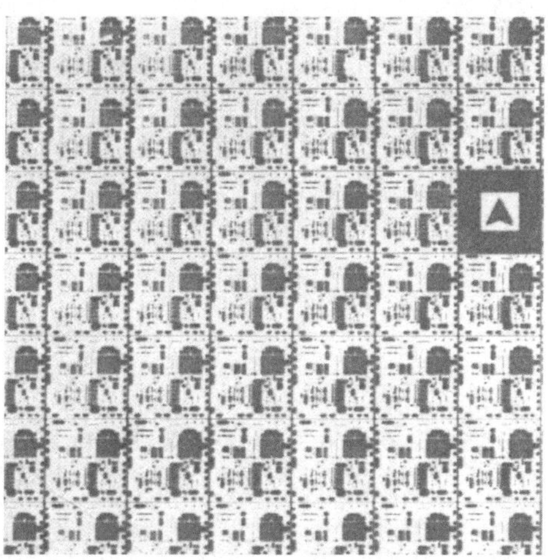

Fig. 1. Part of a 2 inch wafer

In manufacturing an integrated circuit the single crystal rod of silicon is sliced into disks about 0.5 mm thick. After growing of an oxide film on the surface, the surface is coated by a photoresist and exposed to the disired image. The unexposed port of the resist is removed, the unprotected oxide pattern is etched, and the exposed resist is removed. The next operation is vapour diffusion of the doping elements etc. to form the devices. Oxidation, photolithography, etching diffusion and so on can be repeated to build up the circuits.

However, we should bear in mind that trace elements are not all impurities in this sense. The controlled localized addition of elements, as in the diffusion step, is the means by which the devices are produced.

The doping levels are of the order of $10^{19}$to 10^{18} atoms per cm^3. An electrical effect is already observed in a range of

10^{18} atoms/cm^3 Na
10^{17} atoms/cm^3 Cu
10^{16} atoms/cm^3 Au

which is close to the detection limits of the most sensitive analytical methods known.

To allow this control to be exercised, we need starting material (blanks) 100 to 1000 times better than this. This means less than about 10 ppb and it must be the total for all electrically active impurities. A usual criterion for purity is resistivity, which is related to the number of carriers available. But a resistivitiy measurement does not give unequivocal results. Energy levels are similar for many elements, making identification difficult. The compensation or interaction of two carriers may obscure the presence

[1]) Address: 7050 Leipzig, Permoserstraße 15

*) Extended version of the lecture paper given at the 2nd Working Meeting „Radioisotope Application and Radiation Processing in Industry" 28th September to 1st October 1982 Leipzig

of some impurities. Many impurities (O_2) are not active under all conditions and they may be dormant. However, the determination of the presence or absence of impurities is most desirable.

In this context nuclear analytical methods are gaining importance. Reasons for this are that

- radioactive elements can be detected relatively easily with high sensitivity:
- atoms (isotopes) of one and the same element can be distinguished. This results in further advantages so that processes which cannot be observed with other methods can be studied by means of radionuclides.
- for kinetic studies (diffusion, adsorption etc.) the expenditure of time and material is frequently very much lower than with conventional methods.

The most important radiochemical methods of analysis which we use are activation and radiotrace analysis and in some cases the substöichiometric isotope dilution analysis. From the point of view of the analytical problems our work is mainly centred round bulk analysis (quantitative element determination referred to volume) lateral distributions (distribution of a trace on surfaces) depth distribution (diffusion profiles, implantation) interface kinetics and distribution (relation of boundary surface).

Fig. 2 shows the different analytical tasks in semiconductor research and industry.

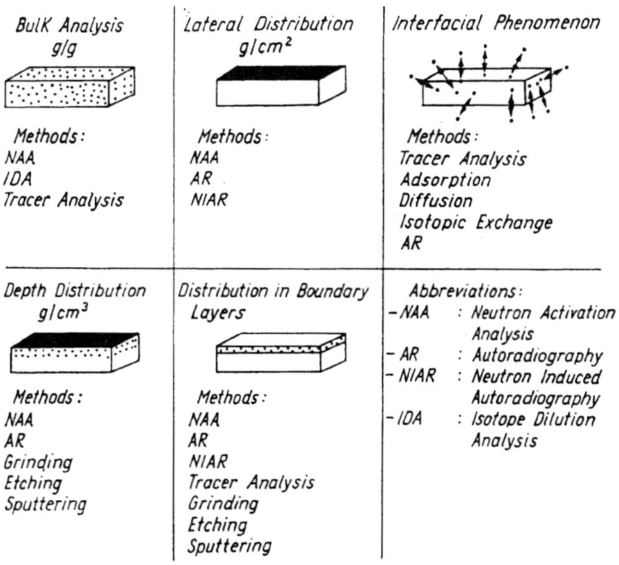

Fig. 2. The important radioanalytical task in microelectronik-processing

Bulk analysis

An example of bulk analysis done by radiochemical methods of trace analysis is the determination of impurities in Si and Se semiconductor material by activation analysis.

The most important radiochemical methods of analysis for content determination (determination of tiny amounts) are activation and isotope dilution analyses.

If a sample is irradiated with neutrons or charged particles the activity produced, A, is

$$A = \lambda N^* = \varphi\sigma N(1 - e^{-\lambda t})$$

and, after irradiation time T

$$A_T = \varphi\sigma N e^{-\lambda T}(1 - e^{-\lambda t})$$

After very long irradiation (as compared with the half-life) the activity produced, A_∞, is proportional to the particle flux, φ, the activation cross-section, σ, and the number of atoms to be determined, N:

$$A_\infty = \varphi\sigma N$$

with λ = decay constant, N^* = number of activated nuclei.

The detection limit depends mainly on the neutron flux and the cross-section.

Besides neutron activation, activation with charged particles is becoming more and more importand for the determination of C, N, O, etc. in the semiconductor industry. There the following nuclear reaction are of particular interest.

$^{12}C(d,n)^{13}N$ (detection limit $\approx$0.1 ppm)
$^{12}C(p,\gamma)^{13}N$
$^{12}C(p,d)^{11}C$ (threshold energy, 18 MeV)
$^{12}C(p,pn)^{11}C$ (threshold energy, 18 MeV)
$^{14}N(p,\alpha)^{11}C$ (detection limit $\approx$1 ppm)
$^{14}N(d,n)^{15}O$
$^{16}O(d,n)^{17}F$
$^{16}O(t,n)^{18}F$

Interferences

In activation-analytical practice some factors which disturb the analysis or give rise to false results have to be taken into account. These include the undesired (disturbing) nuclear reactions. Two kinds of interferences (n, p); (n,α) reactions and second order reactions occur.

For example, in the determination of As, Sb, Br, Te in high purity selenium by neutron activation these processes take place in addition to the (n, γ) reactions:

$^{76}Se(n, p)^{76}As$
$^{79}Br(n, \alpha)^{76}As$
$^{127}T(n, p)^{127}Sb$

Hence direct determinations($<10^{-6}$ %) of the above elements are impossible or are greatly falsified.

A neutron activation analysis for phosphorus in high-purity silicon yields an erroneously high P content due to $^{30}Si(n, \gamma)^{31}Si \xrightarrow{\beta} {}^{31}P(n, \gamma)^{32}P$.

Side-reactions also occur in reactions with charged particles, e. g. in the carbon determination based on the nuclear reaction $^{12}C(d, n)^{13}N$. These side-reactions also yield ^{13}N, thus giving too high carbon contents:

$^{14}N(d, t)^{13}N$ (threshold energy, 4.9 MeV)
$^{14}N(d, dn)^{13}N$ (threshold energy, 12.0 MeV)
$^{16}O(d, 2n)^{13}N$ (threshold energy, 8.4 MeV)

Examples of bulk analysis

As an example of the application of activation analysis by the determination of impurities in semicunductor material our work on Si and Se matrices can be mentioned.

Tab. 1. Trace-element determination in Si by neutron-activation-analysis. Detection limit $3\sqrt{n}$. Average values for a 2 inch disk n-Si ⟨111⟩ [atoms/cm³]

Trace	*Czochralsky* material	zone floated material
Na	$(3.43 \pm 0.20) \cdot 10^{13}$	$(2.92 \geq 0.16) \cdot 10^{13}$
Cu	$(2.22 \pm 0.27) \cdot 10^{13}$	$<(1.5 \pm 0.3) \cdot 10^{13}$
Ag	$(1.17 \pm 0.3) \cdot 10^{14}$	$(1.8 \pm 0.4) \cdot 10^{13}$
Fe	$<(6.5 \pm 2.0) \cdot 10^{15}$	<D.L.
Ni	$<(1.4 \pm 0.4) \cdot 10^{15}$	<D.L.
Zn	$<(1.2 \pm 0.2) \cdot 10^{14}$	<D.L.

Tab. 2. Detection limits for neutron activation analysis and for radioelements in the carrier-free form (detection limit 3·7 Bq)

Isotope	$t_{1/2}$	Common NAA	NAA calculated*)	Carrier-free	Dependence on the sample amount (2 inch Si slice)	
					Disk 100 µm (0·47 g)	Layer 10 nm (0·05 mg)
^{24}Na	15·0 h	10^{-10}	$2 \cdot 10^{-13}$	$1{\cdot}2 \cdot 10^{-17}$	$8{\cdot}3 \cdot 10^{-9}$	$1{\cdot}2 \cdot 10^{-13}$
^{32}P	14·3 d	10^{-9}	$2 \cdot 10^{-12}$	$3{\cdot}5 \cdot 10^{-16}$	$1{\cdot}1 \cdot 10^{-9}$	$1{\cdot}5 \cdot 10^{-13}$
^{59}Fe	45·1 d	10^{-7}	$1 \cdot 10^{-9}$	$2{\cdot}0 \cdot 10^{-15}$	$2{\cdot}0 \cdot 10^{-9}$	$2{\cdot}5 \cdot 10^{-13}$
^{60}Co	5·3 d	10^{-10}	$3 \cdot 10^{-12}$	$8{\cdot}9 \cdot 10^{-14}$	$2{\cdot}5 \cdot 10^{-9}$	$3{\cdot}0 \cdot 10^{-13}$
^{64}Cu	12·8 h	10^{-9}	$2 \cdot 10^{-12}$	$2{\cdot}6 \cdot 10^{-17}$	$2{\cdot}3 \cdot 10^{-9}$	$3{\cdot}2 \cdot 10^{-13}$
^{65}Zn	245 d	10^{-9}	$1 \cdot 10^{-11}$	$1{\cdot}2 \cdot 10^{-14}$	$2{\cdot}3 \cdot 10^{-9}$	$3{\cdot}3 \cdot 10^{-13}$
^{76}As	26·6 h	10^{-10}	$8 \cdot 10^{-14}$	$6{\cdot}4 \cdot 10^{-17}$	$2{\cdot}7 \cdot 10^{-9}$	$3{\cdot}7 \cdot 10^{-13}$
^{110m}Ag	270 d	10^{-9}	$8 \cdot 10^{-12}$	$2{\cdot}3 \cdot 10^{-14}$	$3{\cdot}9 \cdot 10^{-9}$	$5{\cdot}4 \cdot 10^{-13}$
^{122}Sb	2·8 d	10^{-10}	$0{\cdot}9 \cdot 10^{-13}$	$2{\cdot}6 \cdot 10^{-16}$	$4{\cdot}4 \cdot 10^{-9}$	$6{\cdot}1 \cdot 10^{-13}$
^{198}Au	2·7 d	10^{-11}	$4 \cdot 10^{-15}$	$4{\cdot}1 \cdot 10^{-16}$	$7{\cdot}1 \cdot 10^{-9}$	$9{\cdot}8 \cdot 10^{-13}$

*) Calculation depents on ideal conditions: Neutron flux $2 \cdot 10^{14}$ cm^{-2} s^{-1}; Sample amount 0.5 g; Time of activation 24 h; Without interfering reactions; Counting time 64 h

The trace determination in Si by NAA involve no trouble because of the relatively small matrix effect. In most cases up to 20 elements can be measured simultaneously by coincidence spectrometry without chemical separation or only by group separation. In the literature data processing by means of computers has become highly desirable.

Tab. 1 shows some results on *Czochralsky* and zone floating silicon. As you can see the amounts of impurities in zonefloated Si are generally smaller than in *Czochralsky* Si. The detection limit of NAA does not only depends on n-flux, interfering reactions and so on but also on the flux distribution, flux depression, size and amount of the sample.

The influence of the amount of the sample is shown in Tab. 2.

If we assume that 10^{14} atoms per cm^3 of impurities or dopants alter the electrical properties of the semiconductor materials and the depth resolution must be < 1 µm and > 10 nm the amounts of impurities in a 100 µm thick 2 inch diameter slice (0·5 g) are in the order of 10^{-9} g/and in a 10 nm slice (0·05 mg) they are 10^{-13} g. These values are below the detection limits.

The activation analysis of impurities such as Ag, As, Au, Cd, Ga, Sb, Tl in a selenium matrix was elaborated in our lab. In such a case instrumental neutron activation analysis (INAA) is not applicable because of the high activity of the matrix element and the interferences of other radioelements produced during the activation. Therefore extensive chemical separation processes are necessary. More information on this subject will be given in the paper of Dr. *Rudolph* this afternoon.

Lateral distributions [problems of homogeneity]

An important requirement in preparing semiconductor devices is a clean surface without defects. Most of the problems associated with surface preparation are physical ones. However, some chemical contamination does occur from the etches by (chemicals) acids, from the surrounding heating devices etc. The influence of the oxide layer on the electric structure of the surface depends on the bond of the oxygen atoms and the amount of impurities. An average value on the whole surface can be obtained by counting techniques. In most cases we need more important distribution characteristics, but these can be studied by autoradiography.

Autoradiographic methods

An important method within the framework of radiotracer techniques for the determination of lateral and depth distributions is provided by autoradiography. The radioactive-labelled objects under investigation are brought into close contact with halogen-silver emulsions or so-called solid-state track detektor foils. The effect of the ionizing radiation becomes apparent in photo-emulsions as blackening and in solid-state track detectors as track density. Since there is a linear relation between blackening or track density and substance concentration in the object autoradiograms directly yield relative data on concentration distributions of radioactive substances while after calibration they give even absolute values for local concentrations. The technique is applicable in the microscopic and macroscopic ranges. The local resolving power attainable is in the µm range.

The methods of autoradiography have been elaborated to a high degree, they are practicable and are constantly being developed and complemented according to the require-

Tab. 3. Performance parameters of autiographic methods

Method	Measuring	Local resolving
MaAR	as desired	>20 µm
MiAR	a few cm^2	>1 µm
EMAR	a few cm^2	$10^2 \ldots 10^3$ Å
NIAR	as desired	>6 µm

Detection sensitivity of AR

Radiation detection range for radiation fluence[1])

α-rays	1 particle · cm^{-2}
β-rays	$3{\cdot}7 \cdot 10^7 \ldots 3{\cdot}7 \cdot 10^{-3}$ Bq · cm^{-2}
γ-rays	$3{\cdot}7 \cdot 10^8 \ldots 3{\cdot}7 \cdot 10^{-1}$ Bq · cm^{-2}

Accuracy for radiation fluence measurement in %

Absolute counting ±10%

Relative counting with objects for comparison ±6%

[1]) exposure times 10 s to 4 months

Abbreviations

Ma = macro; Mi = micro; EM = electron microskop; AR = autoradiography; NI = neutrón induced

ments of the semiconductor industry, e.g., with a view to radionuclides which are of particular interest technologically.

Some important parameters of autoradiography are listed in Tab. 3.

The method of neutron-induced autoradiography (NIAR) which was developed only recently also allows the representation of distributions of elements having no radionuclides which are useful for tracer techniques, such as boron and nitrogen. The principle is that unlike activation analysis in neutron irradiation it is not the primarily originating nuclide but the particles formed in secondary nuclear reactions (α,p) which produce latent tracks in a cellulose acetate or cellulose nitrate foil which are then developed and recorded. The principle is shown in Fig. 3. The material to be analysed is activated by a n-source in close contact with the detector foil. The advantage is that β-, γ- and n-radiation does not influence the basic reactions $^{10}B(n,\alpha)^{6}Li$; $^{14}N(n,p)^{14}C$; $^{7}Li(n,\alpha)^{3}H$. The etched foil can be evaluated by a microscope.

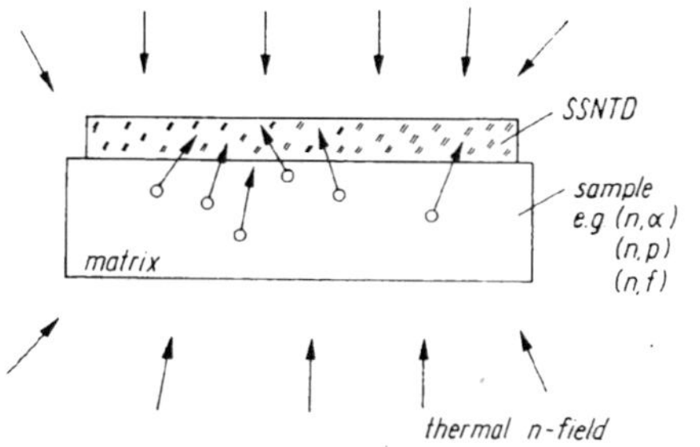

Fig. 3. The principle of neutron induced autoradiography (NIAR) SSNTD = solid state neulcar track detector

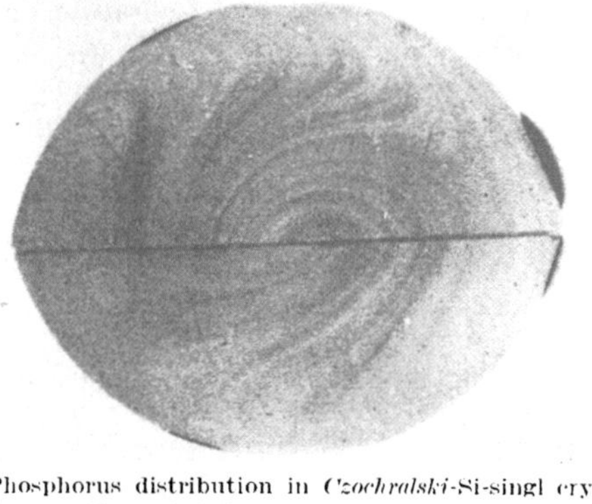

Fig. 4. Phosphorus distribution in *Czochralski*-Si-singl crystals. P was given to the melt before

Some parameters of neutron-induced autoradiography as used for boron and nitrogen determinations are found in Tab. 4.

Fig. 5. Inhomogeneous P-distribution on a Si disk with circuit structure after a high temperature diffusion process

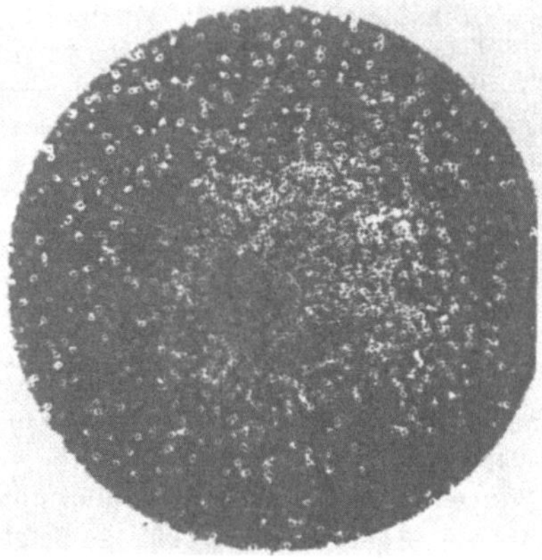

Fig. 6. Solid state track detector autoradiogram of an inhomogeneous B distribution in silicon after diffusion

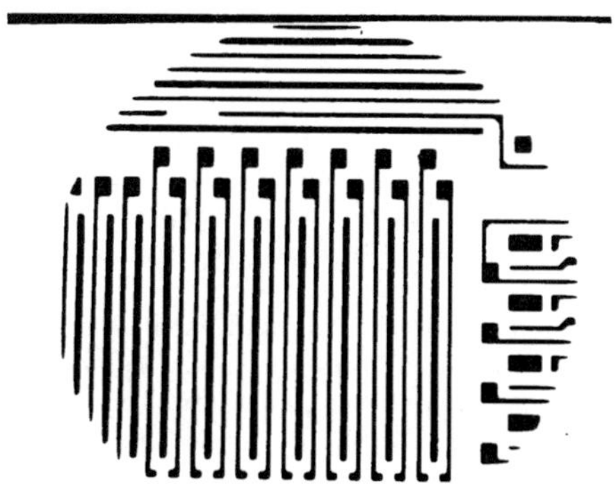

Fig. 7. Photo-mask of a small part of a semiconductor circuit

Tab. 4. Performance parameters of NIAR as used for the determination of B and N

	Boron determination	Nitrogen determination
Resolution	$\geq$ 10 μm with CA 80-15 (CN) $\gtrsim$ 5 μm with T-Cellit (CA) $>$ 0.3 μm with EM	$\geq$ 8 μm with CA 80-15 and ZfI-CN-foil
Sensitivity	$C_B = 10^{13} \ldots 10^{22}$ atoms $^{10}B\ cm^{-2}$ (reactor, neutron fluence $10^{11} \ldots 10^{14}\ n_{th}\ cm^{-2}$) $C_B = 10^{15} \ldots 10^{22}$ atoms $^{10}B\ cm^{-2}$ (3 mg ^{252}Cf-source; t_{exp} 30 d)	$C_N \leq 0.5$ wt.-% ^{14}N (reactor and Cf source neutron fluence $10^{12}\ n_{th}\ cm^{-2}$)
Detection limit	$\geq 10^{15}$ atoms · cm^{-3}	$> 10^{21}$ atoms · cm^{-3}
Accuracy	±10% for $C_B = 10^{16}$ atoms · cm^{-2} (reactor) ±50% for $C_B = 10^{16}$ atoms · cm^{-2} (Cf-source)	

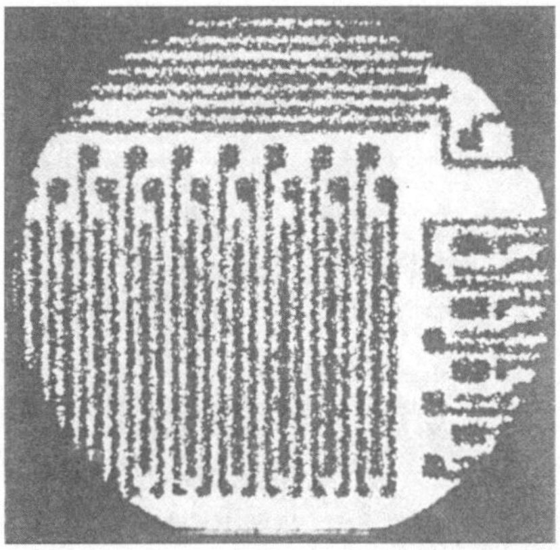

Fig. 8. Boron solid state track autoradiogram of the circuit part shown in Fig. 7

A few examples may serve to show the efficiency of autoradiographic methods. Fig. 4 gives the phsophorus distribution in Cz-Si single crystals. The phosphorus was uniformly distributed in the melt before. An inhomogeneous distribution with enrichment at the phase interfaces which may also be due to uneven solidification is found.

Since dislocations and stacking faults exhibit a pattern similar to the P distribution studies of these relationships are highly interesting. Fig. 5 presents the P distribution and surface impurities on a Si disk with a circuit structure as may arise in the course of the manufacturing process.

Fig. 6 shows a solid-state track boron distribution from a boron determination in a Si disk after high-temperature boron diffusion. Here again an inhomogeneous distribution is seen.

The resolving power in NIAR which we have attained so far is demonstrated in the Fig. 7 and 8. Fig. 7 shows the photo-mask of a strongly enlarged part of a circuit. The real diameters of the conductions are 3 μm. Fig. 8 gives the corresponding boron solid-state track radiogram.

One can say that contrary to instrumental surface analysis methods such as SIMS, ESCA, AES etc. Autoradiographic methods allow us to analyse large surface areas in a simple way. The lateral resolution is in the range of μm and one can get a survey of the distribution of impurities on large surfaces very quickly.

Investigation of depth distributions

For the development of VLSI circuits with short switching times and sharply defined n-p junctions by doping very small surface bands the depth distribution must be known and controlled by analytical methods. On the other hand the quality of devices depends on the distribution of mobile charge carriers, e.g. in dielectric films such as SiO_2 and Si_3O_4.

Depth distributions of dopants or impurities in silicon disks must be determined step by step after labelling (e. g., by activation in a reactor, by anodic oxidation, cutting, polishing, etching or sputtering and subsequent activity measurements). These techniques are very sophisticated; the accuracy and resolving power depend on the efficiency of the particular method of removal. For several years we have made such studies of depth distribution using activity measurements, autoradiography and solid-state track reactions for the determination of elements such as B, P, Au, Na etc., the attainable depth resolving power being $\geqq 10$ nm.

In what follows a few examples of our work concerning the determination of depth distributions in semiconductor components by means of activation analysis will be given. Methods of ion implantation are being used increasingly in large-scale integrated circuits with accurately defined n-p junctions. They have the advantage over thermal diffusion that they allow charge transitions in narrow surface layers. For the development of optimum implantation techniques it is necessary to know the connection between ion energy, penetration depth and dopant profile. Such depth profiles can be determined by activating the implanted disks and removing layer after layer (anodic oxidation) with radiochemical (activation-analytical) content determination. In Fig. 9 a depth profile of implanted boron in silicon determined by NIAR is presented. Fig. 10 shows the depth distribution of phosphorus in CVD system polysilicon, SiO_2, Si after chemical etching and counting and Fig. 11 shows the distribution in the same system but done by sputtering and autoradiography. A comparison shows that both results are identical.

An important problem is the analysis of dielectric films such as SiO_2 and Si_3N_4. Activation analysis is the only technique with sufficient sensitivity to analyze these films and obtain information on the distribution of impurities through the film. Our work is restricted mainly to determining sodium in SiO_2 films on silicon.

Analyses for sodium in SiO_2 films on silicon became important in the semiconductor industry with the introduction of MOS technique.

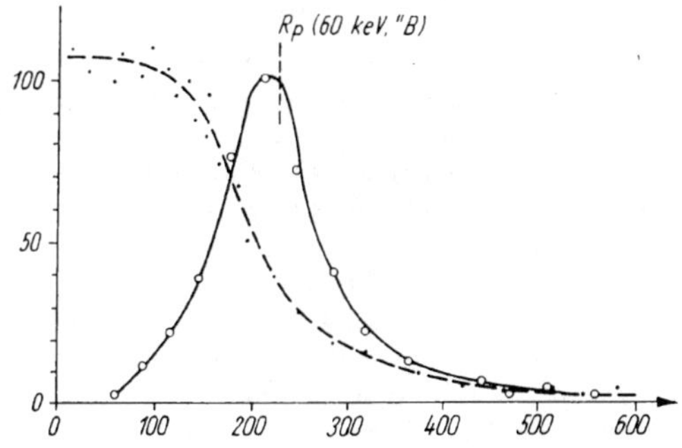

Fig. 9. Depth profile of implanted boron in silicon determined by NIAR

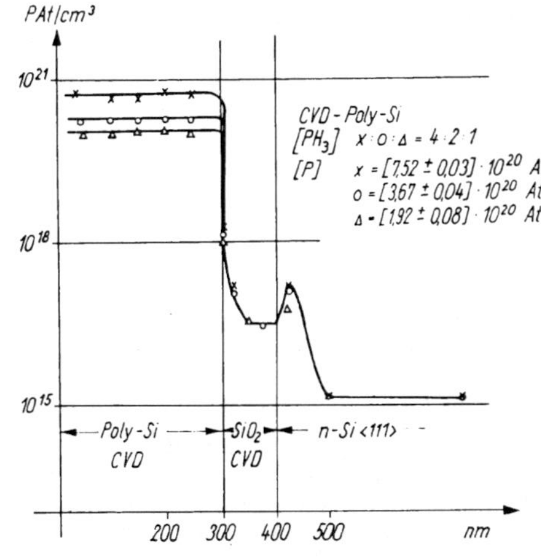

Fig. 10. Depth distribution of phosphorus in CVD-system polysilicon, SiO_2, Si determined by chemical etching

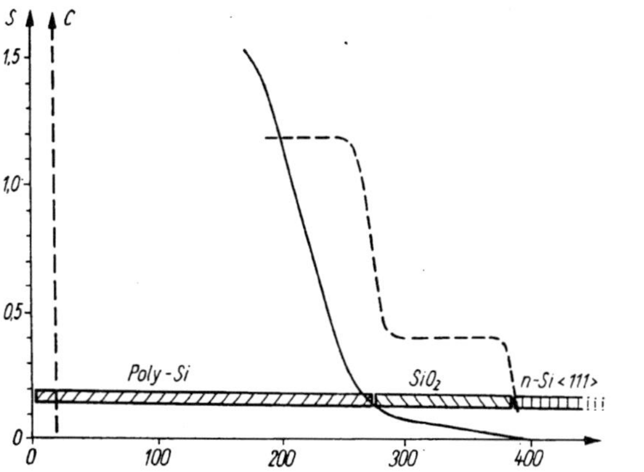

Fig. 11. Depth distribution of P in CVD-system polisilicon, SiO_2, Si determined by sputtering and NIAR

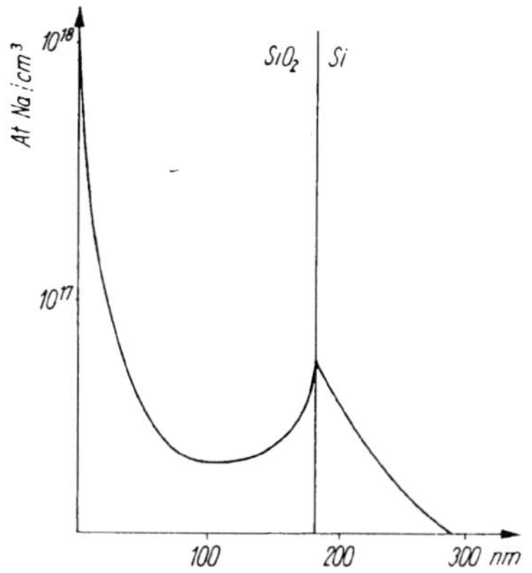

Fig. 12. Sodium distribution in the system SiO_2, Si after gate oxidation

In most cases the oxide is used as a insulator to passivate a device. But in MOS field-effect transistors e.g. the oxide film becomes an active part of the devices. In such a case the oxide film supports a thin metal film (gate) which is used to supply charge through the oxide to the substrate. The metal oxide acts as a gate similar to a grid in a vacuum tube. To prevent instability of the device the silicondioxide film must be virtually free of any mobile ions which can drift through the oxide if a bias is applied to the gate. Because the high mobility and aboundance sodium ions has been shown as the principle sources of such mobile ions in these films.

A typical sodium impurity profile through an oxide film is shown in Fig. 12. After irradiation and rinsing in 6 N hydrochloric acid the oxide film is then incrementally etched in 50 to 500 Å (5—50 nm) steps using 5% HF and each lap is analyzed radiochemically. One can see, the oxide getters the sodium.

Radioactive tracers

In the characterization of semiconductor materials, oxide films, etc. for chemical imperfections, it is sometimes more convenient to use radioactive tracers than activation analysis. There are generally two reasons for choosing this approach. Firstly, the chemical impurity may not yield a suitable species (half-life too short, poor yield due to small capture crossection or very long half-life, or the radioactive species does not lend itself readily to counting).

Secondly, there may be a predominance of other impurities, yielding radioactive species that obscure the activity of the impurity of interest.

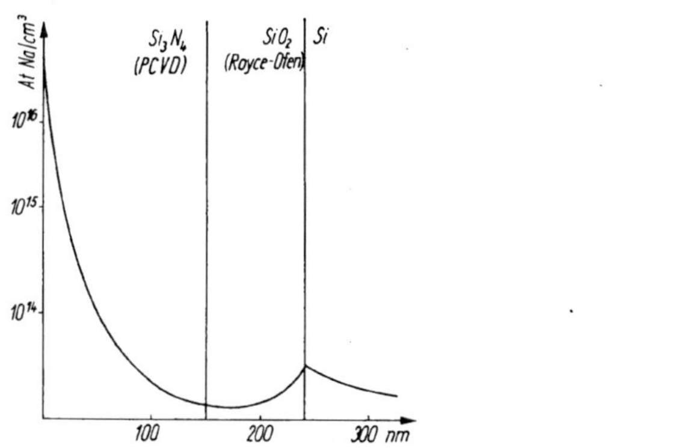

Fig. 13. Sodium-22 distribution in a silicon nitride silicon dioxide layer after tempering for 30 min at 850 °C

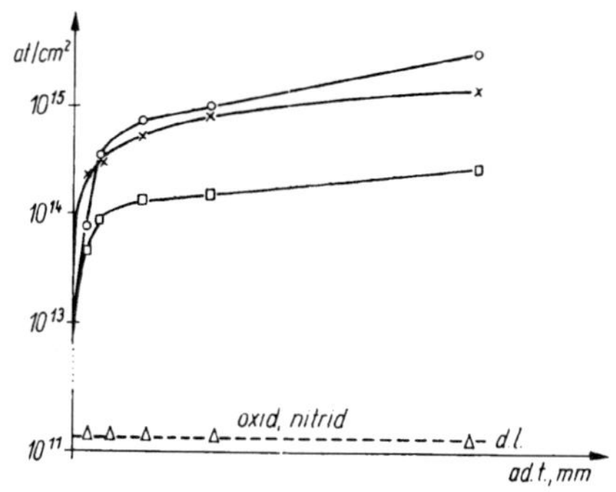

Fig. 14. Ag-adsorption in Si, Si_3N_4 and SiO_2 surfaces
The adsorption on Si_3N_4 and SiO_2 is smaller than the detection limit.

$[Ag^+] = 7{\cdot}5 \cdot 10^{-6}$ mole/l; ○ = acetone, × = etcher V; □ = HF (38%); △ = adsorption on oxide, nitride

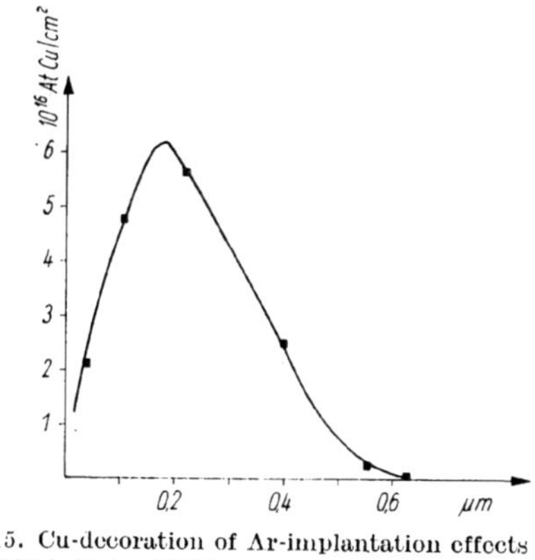

Fig. 15. Cu-decoration of Ar-implantation effects of Si p-Si, 260 keV Ar, 10^{16} ions cm^{-2}, 30 min 600 °C

An example of such a short-lived activity is the study of sodium in Si_3N_4 and SiO_2 films. In the n-activation analysis of sodium only ^{24}Na (half-life, 15 h) is produced. The short half-life precludes longer experimental work, because all available time is used in the analysis. However, sodium has a long-lived radioactive isotope, sodium-22 (2·6 years) that can be used as a tracer.

Fig. 13 shows a typical sodium concentration profile through a silicon nitride, silicon dioxide film (2500 Å) after tempering for 30 minutes at 850 °C. As we can see the sodium migration through the film is very small. The nitride and the dioxide getter the sodium.

One of the most important requirements in preparing semiconductor devices is a clean surface without defects. Besides the physical problems associated with surface preparation, also chemical contamination occurs, often from the etching, diffusion and CVD processes. In such cases the source of such impurities must be identified. Fig. 14 shows the adsorption of Ag^+ ions on silicon surfaces and on insulator layers. The adsorption out of a 10^{-6} mol per litre solution is about 10^{14} atoms/cm^2 and not negligible.

The last Fig. 15 shows the decoration of implantation defects by Cu^+ ions. There is a relationship between defects and Cu^+ to be adsorbed.

In summary it can be said that nuclear analytical methods have become of paticular importance for the analytical control of MOS base materials, for the development and monitoring of technological steps or processes and for the checking of final products in the semiconductor industry.

What is determined is, besides contents, especially lateral and depth distribution of trace elements in the ppm to the ppb region. In these regions it is not enough to use one method elaborate as it may be but one must employ reference methods and methods involving high-precision standard samples and ring analyses. This is an extremely expensive and time-consuming procedure, so the expenditure and accuracy required in a particular case have to be carefully considered.

From this point of view analyses with tracer techniques can be regarded as a useful and necessary complement to the known trace and surface methods since their cost is low.

In future trace analysis may be an interesting tool to study the relationship between physical properties, physical structure (offsets, swirls etc.) and the distribution of dopants and impurities. The major challenge to trace analysis now is analytical control during the manufacturing of VLSI devices.

Received March 31, 1983

Isotopenpraxis 20 (1984) 2, S. 47 bis 52

Originalarbeiten

Axial Dispersion and Homogenization in Chemical Engineering Systems*)

A. Merz

(Kernforschungszentrum Karlsruhe GmbH, Laboratorium für Isotopentechnik)[1]

The knowledge of process streams is an essential assumption for the optimization of apparatus and methods in chemical industry. The application of radiotracer methods for defining the intensity of axial mixing and homogenization effect is demonstrated by means of three selected examples of technical systems: powder mixing, waste water treatment, chemical process in a rotary kiln.

Die Kenntnis der Prozeßströme ist eine wesentliche Voraussetzung für die Optimierung von Apparaten und Verfahren in der chemischen Industrie. Die Anwendung von Radiotracerverfahren zur Ermittlung der Intensität der axialen Vermischung und der Homogenisierung wird an drei ausgewählten Beispielen technischer Systeme gezeigt: der Pulvermischung, der Abwasserbehandlung und des chemischen Prozesses in einem Drehofen.

Keywords

data processing; mathematical models; mixing; response functions; tracer techniques

1. Introduction

The knowledge of the motion characteristics of the process flows in the chemical industry is a basic requirement for the optimization of equipment and processes. A large volume of detailed information about the local dynamic behaviour can be successfully obtained by means of radiotracer experiments in relation with multi-detector arrangements. The system of transfer properties of constructive parts, flow channels and branching can be thus evaluated.

A recently introduced-32-Detector Collection System (Fig. 1) controlled by microprocessors for residence time measurements with radionuclides is at the disposal of the alboratory for Isotopic Technology. The system can be autonomously operated, as well as on-line with a PDP 11/60 Process-Computer, which organizes Data Transfer and Data Processing in a multi-user mode.

2. Process Analysis of the Motion Mechanism

It will next be reported about the process analysis of the axial dispersion and homogenization.

Up to now, the afore-mentioned process analysis has been carried out by a Data Acquisition System based on a modified multichannel analyser.

2.1. *Continuous mixing of solids*

The first example is taken from the field of the continuous mixing of powder-like materials (Fig. 2). The ploughshare mixer, in regard to a continuous operation investigation, was provided with the necessary periphery and equipped with 10 scintillation detectors. The process investigation objectives were defined as the description of the influence of different constructive arrangements of the mixing chamber on the solid particle motion [1].

To this, the experimental bruised maize grist was tagged at the mixer entrance with an identical radioactive sample (with the radionuclide ^{24}Na and ^{56}Mn).

Some resulting response functions, after the impulse injection, measured by detectors (1 – 6), are presented on

*) Presented at the 2nd Working Meeting "Radioisotope Application and Radiation Processing in Industry", 28th September to 1st October 1982, Leipzig

[1]) D-7500 Karlsruhe, Postfach 3640

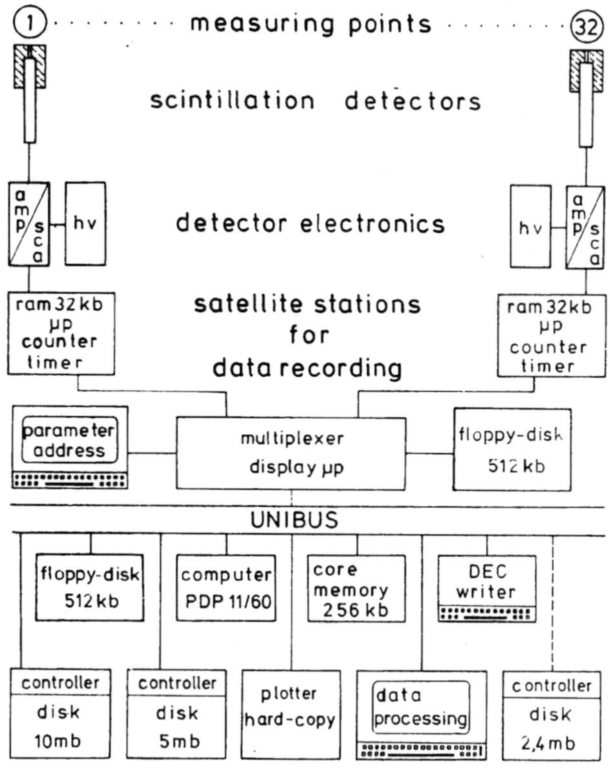

Fig. 1. Computer controlled data recording and data processing

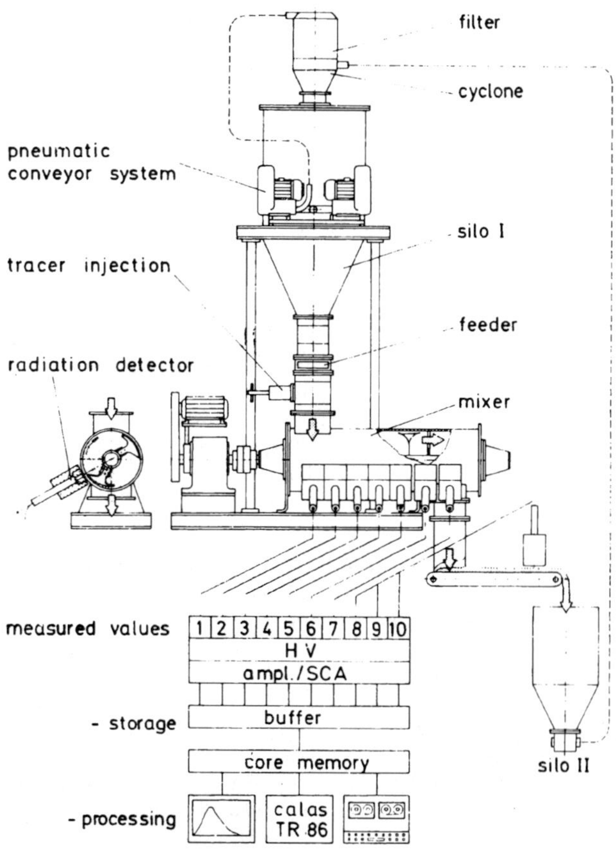

Fig. 2. Schematic representation of the test facility with electronic measuring system

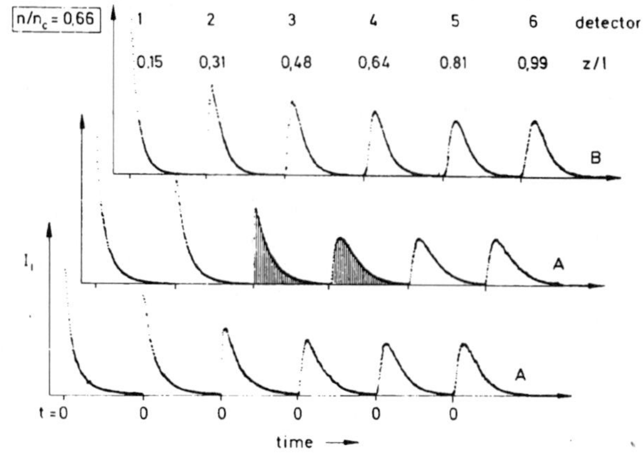

Fig. 3. Number of pulses as a function of time recorded at six levels of measurement

Fig. 3 for the rotation speed of the mixer shaft $n/n_c = 0{\cdot}66$ ($Fr = 0{\cdot}44$). The different geometric arrangements of the cilindrical mixing chamber are represented, as A, A′ and B. A and B are ploughshare configurations which strongly differ. The mixing chamber in A′, opposite to A, is equipped with an additional reflector wall.

The effect of this wall, in relation to the local solid particle motion in the mixer, is elucidated through the raster measure curves and quantitatively evaluated through the calculation of the statistical moments, mean value and standard deviation, of the time function of the tracer concentration [2].

The results of these calculations are shown on Fig. 4 in a normalized form along the axial mixing path z/l.

According to this, and in spite of a slightly smaller hold-up, the mean residence time ($z/l = 1$) is about 15% larger, due to the fact that the grist will be strongly slowed down close to the reflector wall.

Through a combination of the experimental results with the dispersion model [2] and considering the border and initial conditions, the motion mechanisms in continuous mixing can be described by the transport coefficients v (convection velocity) and D (coefficient of axial mixing). Fig. 5 shows the functional development of the normalized convection velocity over the Froude number.

Two regions can be determined according to this:

For $Fr > 4$, the convective velocity is independent from the velocity of the mixing flights.

For $Fr < 4$ the axial transport is approximately proportional to the rotation speed of the machine.

On the other hand, other laws can be deducted for the intensity of the axial mixing (Fig. 6):

For $Fr < 3-4$, the mixing intensity grows proportional to the rotation speed of the machine (A and A′).

In the range $Fr > 4$, the mixing intensity can grow overproportionally ($D \sim n^5$); this can be explained as a result of a drastic reduction of the filling level from 60% to 40% in A and A′. A comparison of the invested power into the blender charge is unavoidable for an overall critic of the mixing process with a variation of constructive arrangements. The corresponding dimensionless specific power is shown on Fig. 7, as *Newton* number Ne against the *Froude* number Fr.

As a rule, no significant influence of the mixing geometry on the mixing power can be determined.

The rotation speed ranges can nevertheless be defined. These ranges can be defined in relation to the power uptake

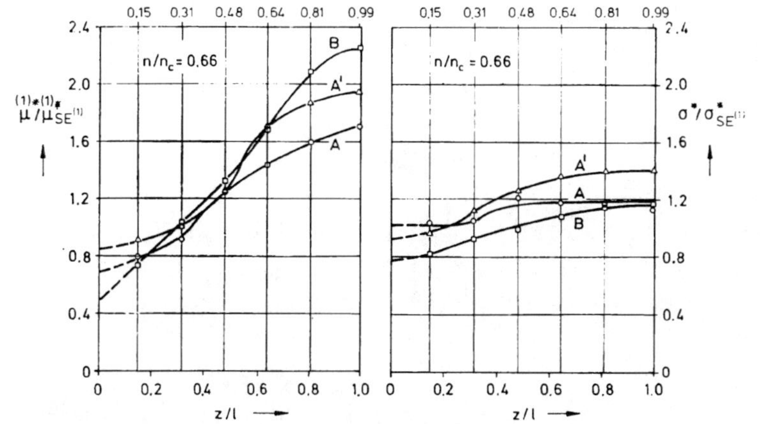

Fig. 4. Statistical moments along the axial mixing path

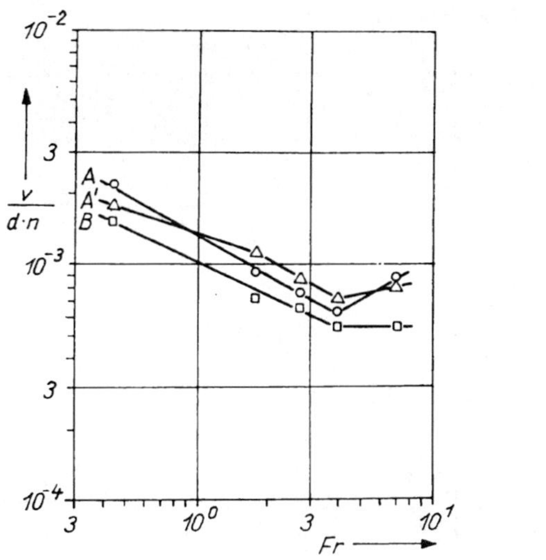

Fig. 5. Convective velocity v as a function of *Froude* number Fr

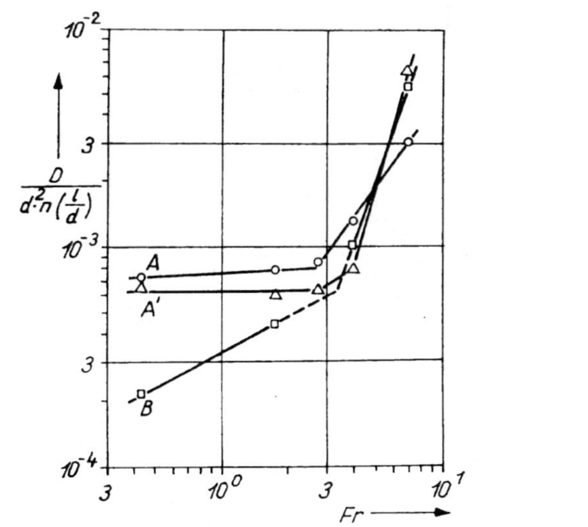

Fig. 6. Axial dispersion coeffizient D as a function of *Froude* number Fr

of the hold-up: the friction effects dominate below $Fr < 1$. There is *Coulomb*-friction between the ploughshare flights and the grist. With a higher rotation speed ($Fr > 1$), the intertia forces of the loose bulk material become increasingly effective: the motion condition of the hold-up is then comparable to that of a fluidized bed.

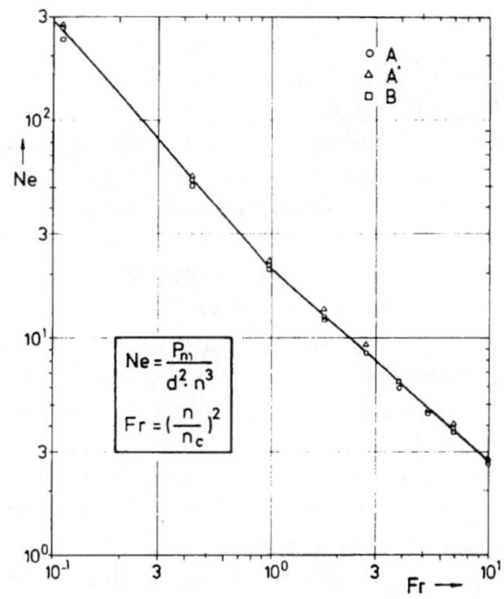

Fig. 7. *Newton* number Ne versus *Froude* number Fr

As a first approach, it can be talked about a ideal-mixed behaviour of the mixer, for $Fr > 7$, under fluid-dynamical consideration of the physical properties, of the experimental stuff.

For a given mixing task, considering the experimental results altogether, recomendations can be made concerning the residence time, residence time distribution and axial dispersion.

2.2. *Fluid Dynamics of aerated wastewater basins*

In municipal and industrial wastewater treatment, minimal residence times are necessary in the contact of dispersed oxygen with the fluid phase and the suspended matter, for the biological degradation of aerobic coloidal and disolved materials. The process investigation objectives were defined, regarding biochemical transformations in wastewater basins, as a fluid-dynamical study the aqueous phase through radiotracer experiments. The basin load (wastewater flow), the graded charge of the basin and the electrical power demand of the simplex stirrer turned out to be decisive parameters for the residence time in the basin, the axial and orthogonal dispersion, as well as dead volume phenomena.

The schematic set-up of the wastewater basin is depicted on Fig. 8. The basin was designed for a capacity of 12000 m^3 and, fitted with 5 simplex turbine stirrers.

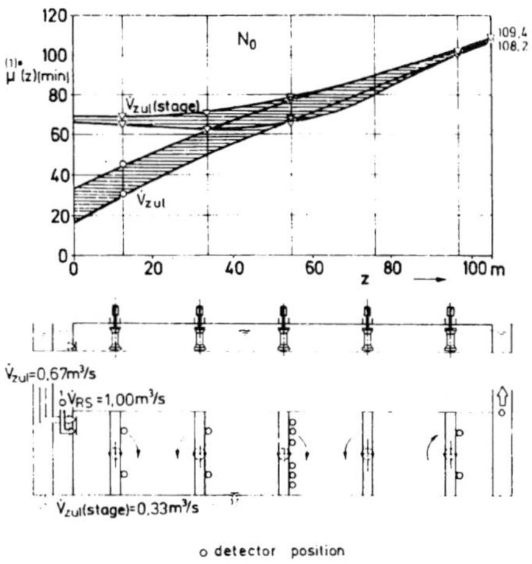

Fig. 8. Influence of stage stream

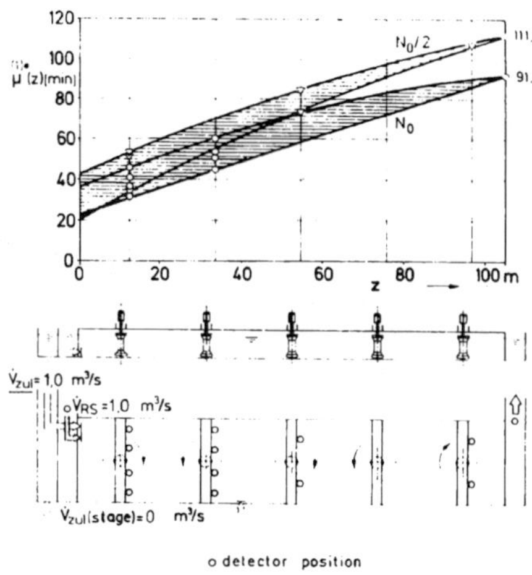

Fig. 9. Influence of stirrer power

The detector positions are shown as circles on the stirrer bridges, drain channel and sludge recycle.

The immersion depth of the detectors in the basin could be varied. In this way, layer streams could be measured.

A sodium-carbonate solution of about 100 mCi ^{24}Na activitiy was used as a tracer for the fluid phase. The tracer was respectively injected as a quasi-impulse-function into the main feed ($\dot{V}_{zul}$) and the stage flow.

The statistical moments as distribution characteristics can be deducted from the time function of the tracer concentration as local response to the feed stream impulse tagging.

It was shown, by comparative experiments, that the mean residence time of the main entrance wastewater flow ($\dot{V}_{zul}$) could be about 49% enhanced as a result of the driving force of a stage stream that was feeded 40 meters downflow. It was also unexpected the fact, that about 50% of this stage stream, due to the backmixing effect of the simplex stirrer, appeared in the main entrance region of the basin (detectors 2/3) already a few minutes after the tracer injection. The mean values $\overset{(1)}{\mu}{}^*(z)$ of the locally measured response functions of both experiments are comparatively shown on Fig. 8. Due to the concordance in the mean values from the middle of the basin downstream to the drain, it could be supposed that the transport mechanisms in the second half of the basin agree, no matter if the fluid elements belong to the main entrance flow or to the stage flow. Clear differences can be however noticed in a detailed analysis of the entire response functions:

- The maximum tracer activity carried by the main entrance stream will be measured in the drain 61 minutes after injection (Detector 10), in comparison with 44 minutes for the stage stream.

These facts have a radical importance for the wastewater technologists in the evaluation of the biochemical oxygen demand of pollutants that affect the water quality, and are decisive for the load operation of such aerated basins.

A second fluid-dynamical experimental comparison took place under energy considerations (Fig. 9). The electrical power demand of the 5 simplex stirrers was reduced 50% from an overall N_0 through a reduction of the rotation speed and torque, keeping all the other experimental conditions unchanged. The decisive distribution parameters were, as usual, calculated from the response functions to the impulse tracer injection into the wastewater stream ($\dot{V}_{zul}$).

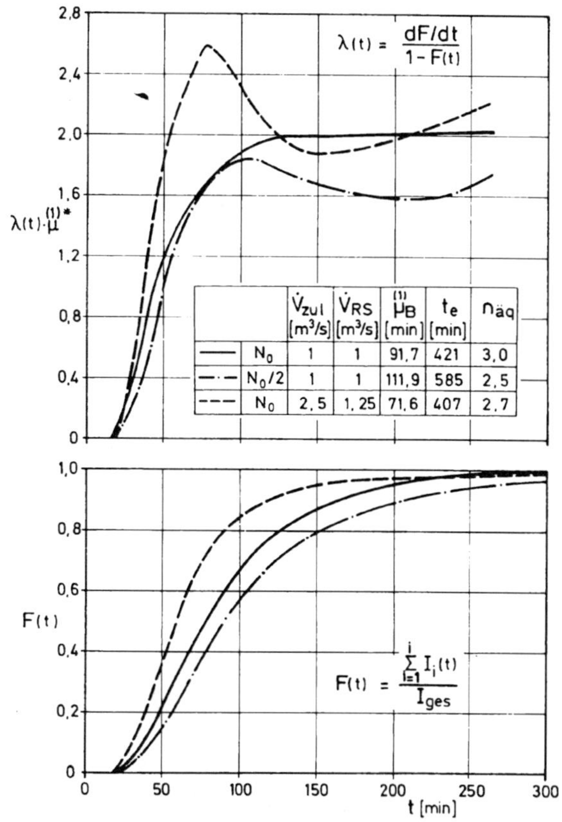

	$\dot{V}_{zul}$ [m³/s]	$\dot{V}_{RS}$ [m³/s]	$\overset{(1)}{\mu}_B$ [min]	t_e [min]	$n_{äq}$
—— N_0	1	1	91.7	421	3.0
—·— $N_0/2$	1	1	111.9	585	2.5
--- N_0	2.5	1.25	71.6	407	2.7

Fig. 10. Dead region analysis: sum function $F(t)$ and intensity function $\lambda(t)$

The functional development of the mean concentration values $\overset{(1)}{\mu}{}^*(z)$ are shown on Fig. 9 in dependence on the corresponding detector position in the basin and the electrical power demand as parameter. The fluid motion in axial direction, as well as perpendicular to the main stream flow, is showed down by a 50% power reduction. The middle fluid elements have a 22% longer residence time in wastewater basins. The cross-mixing is not so intense in the region of the first stirrers, but the concentration distribution is almost homogeneous after around 95 meters.

The minimum residence time was 8 minutes longer, and the activity maximum was measured 13 minutes later as in the comparative experiment with full stirring power N_0.

It was almost obvious to develop a deeper fluid-dynamical judgement basis through a dead region analysis [3]. This analysis was grounded on a 39% longer maximum residence time with tailing effects at reduced stirrer power.

The sum function F(t) and intensity function $\lambda(t)$ are shown on the following diagram (Fig. 10) for both discussed experiments, and for a third experiment with an extreme large basin load. From this follows that no dead region effects are noticeable in the basin operation with full stirrer power N_0 and standard feed flow $\dot{V}_{zul} = 1\ m^3/s$.

The intensity function has, during the considered time, a continuous positive slope.

On the other hand, a small dead region tendency exists by reduced stirrer power. However, this would be acceptable as far as the aeration of the wastewater brings enough oxygen.

When the feed stream is extremely increased, so that enhanced short circuit effects can take place, the fluid dynamics will be decisively modified in accordance to the expectations.

In a following experiment the overloading of the basin was stimulated. A clear-out maximum was now obtained for the intensity function after around 75 minutes. A delayed transport phase from badly mixed basin regions linked up with this maximum. The reported experimental results are part of a larger research program. This program includes, together with the fluid-dynamical residence time distribution analysis, also biological measurements carried out by wastewater technologists.

An overall evaluation is not planned at this moment.

2.3. *Transport mechanisms in a rotary kiln*

The third example for process flow investigations by application of radioactive tracers is taken from the fertilizer industry. The knowledge of the rheological behaviour of reaction mixtures in the production of Rhenania-phosphate by thermal sintering of raw phosphate, sand and soda in rotary kilns, is important in regard to the chemical transformations and a rational process control.

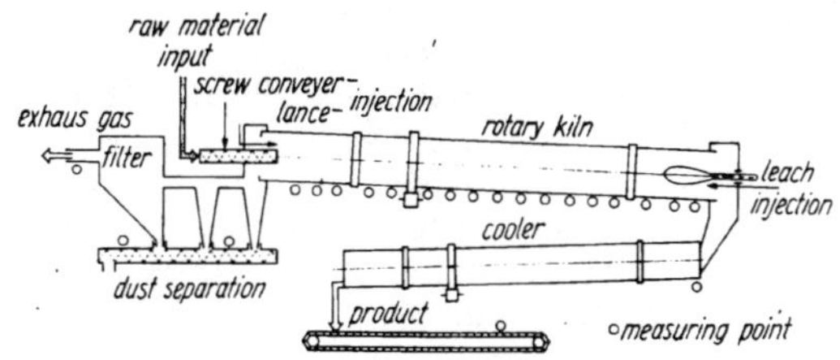

Fig. 11. Schematic representation of fertilizer plant

The rotary kiln (Fig. 11) is a mixer as well as a chemical reactor [4]. The raw feedstock is carried into the first section of the kiln by a screw conveyor, mixed by upheaval in the inner wall of the kiln and preheated by the exhaust gases of the flame heater. The now viscous material reaches the actual sintering zone in the high temperature region of CO_2-degassing where the transformation into Rhenania phosphate takes place.

From the extensive and worked-out research program, three crucial questions should be presented:

- Division of the rotary kiln into process zones by means of residence time distribution analysis in the process components.
- Determination of maximum dust emission zone of the kiln.
- Motion mechanism of the kiln filling by partial substitution of soda through soda lye.

According to the given positions on the facility diagram (Fig. 11), the production unit was fitted for the research project with a total of 20 interchangeable detectors. All of the outgoing flows — product, process dust, exhaust gases — could be, in this way, technically recorded.

Basic experiments for the division of the kiln into process zones were undertaken by means of tagging the components with activated samples of soda, recycle process dust and glass powder simulating the feedstock sand. The applied ^{24}Na activity amounted about 20 mCi every time. It could be shown that the transport mechanisms of these indicators are generally identic in the scope of the acceptable error range for such large experiments. The relative statistic moments mean and standard deviation of locally measured time functions of the tracer concentration — shown on Fig. 12 along the dimensionless reactor length — as well as the measure curve integral I(z) are judgement criteria. The corresponding moments at the reactor exit $z = 1$ are normalizing values for the mean and standard deviation.

The efficiency of every detector arrangement was determined by previous calibration measurements in the "cold kiln". In this way it was assured that the individual signals were comparable to each other. The rotary kiln can be

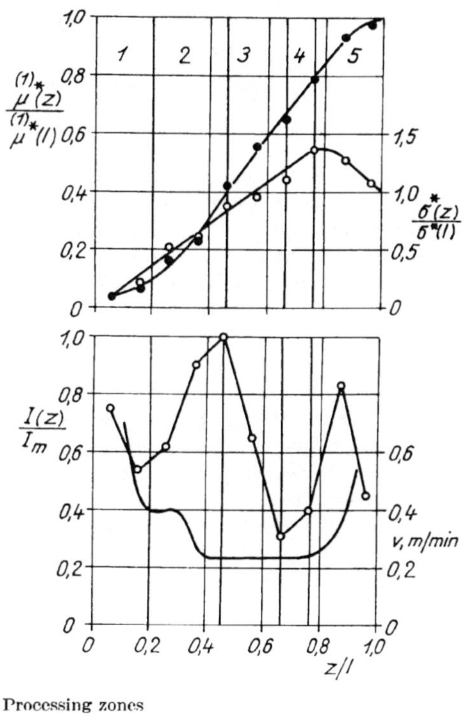

Fig. 12. Processing zones

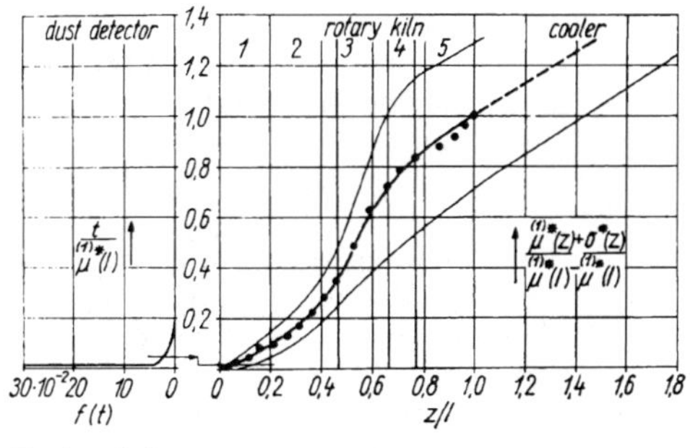

Fig. 13. Dust emission zone

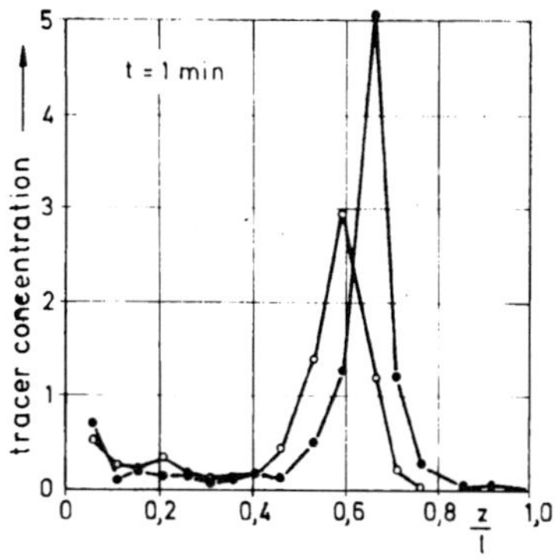

Fig. 14. Tracer concentration distribution of ^{24}Na labelled leach

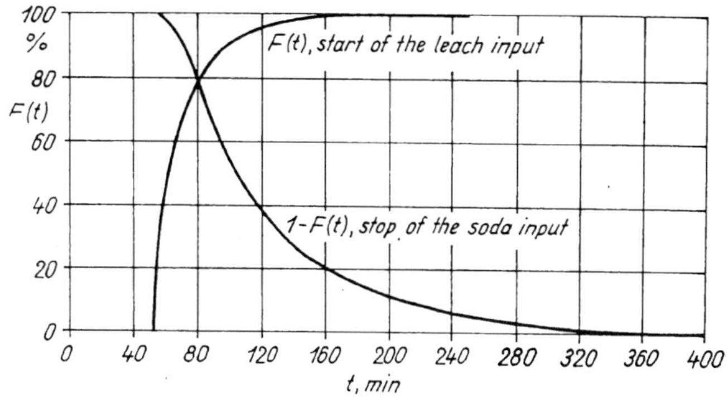

Fig. 15. Transition behaviour: soda stop — leach start, measuring point $z/l = 0{\cdot}71$

divided into 5 process zones through the function development of the specified characteristic dispersion values, and under inclusion of the calculated mean transport velocity v of the indicator. These five zones were also mentioned by *Schügerl* et al. after their experiments with a laboratory rotary kiln [5].

The warm-up zone (2), the CO_2-degassing zone and the reaction or sintering zone (5) could be the most important regions worth to be mentioned. A more detailed interpretation of the relations of the judgement values must now be sacrificed. The response signals of those detectors that were installed at the screw conveyor, behind the dust separator, were decisive, in relation to the question of the lay-out, to reduce the process dust emission.

The residence time distribution as time spectrum of the tracer activity, measured by a dust detector, is shown on the left diagram on Fig. 13. The spectrum can be correlated with the materials' motion in the rotary kiln (right diagram). This motion is once again characterized by the functional development of the statistical moments mean and standard deviation. The residence time in the dust separators was determined with additional lance-injections of radioactive tagged dust and correspondingly corrected. According to these results, process dust is definitely produced in the first zone — the entrance and starting zone — which will be carried by the combustion gases out of the kiln. A significant dust emission in the CO_2-degassing zone can be therefore excluded.

The experimental results give, altogether, important directions for process modifications in order to reduce the dust emission. Other motion mechanisms, in comparison to the kiln operation with solids, must be considered in the partial substitution of soda through soda lye, which will be transported counterflow to the kiln filling. The analysis of this problem is important for the automatic process control during a feedstock change, from pure soda to partial soda-lye operation (and vice versa), with the condition of a constant product quality. For this, the lye stream was radioactively tagged over a by-pass line and its dispersion over the product bed measured and followed as a time function.

Concentration profiles along the kiln's length are shown of Fig. 14, one minute after injection. Both profiles correspond to different substitution relations. The maximal concentration zones are fluiddynamically interesting. The larger lye volume will be brought there into the bed, and therefore, the motion characteristics are of special interest.

On the next figure (Fig. 15), the sum function $F(t)$ of the tagged-lye activity is shown in an experiment with maximum profile at $z/l = 0{\cdot}71$.

It characterizes the expected time development of the lye front at this place. The expected reduction in the sod content of the bed, after switching off the raw soda input, is comparatively shown on the same figure.

With the knowledge of both functions it is now possible to effect the substitution without affecting significantly the product quality.

Symbols

A, A′, B	Different geometric arrangements of the mixer chamber
Bo	*Bodenstein* number
D	Axial dispersion coefficient
$F(t)$	sum function
Fr	*Froude* number
I_i	number of pulses accumulated per measurement time interval
I_{ges}, I_m	total number of pulses
N_0	power
P_m	specific power (in relation to the hold-up)
$\dot{V}_{RS}$	sludge back flow
$\dot{V}_{zul}$, $\dot{V}_{zul}$ (Stage)	water-water input
d	diameter of the mixer
$f(t)$	residence time distribution (RTD)
l	length of the mixer chamber, — of the aerated basin —, of the rotary kiln
n, n_c	speed of the mixer shaft, critical speed
$n_{äq}$	equivalent number of stirred vessels
t, t_e	time, end of RTD
v	convective velocity
z	path coordinate
$\lambda(t)$	intensity function
$\overset{(1)}{\mu}{}^*$, $\overset{(1)}{\mu}{}^*_{SE}$	mean value (weighted average), SE = standard experiment
$\overset{(1)}{\mu}_B$	residence time in the aerated basin
σ^*	standard deviation

Received February 1, 1983
Accepted in revised form May 25, 1983

References

[1] *A. Merz, R. Holzmüller*, IChemE Symposium Series No. 65 (ISBN 085295 134 5)
[2] *O. Molerus*, Chem.-Ing.-Techn. **38** (1962) 137
[3] *E. Otto, W. Gestrich*, Chemie-Technik **4** (1975) 43
[4] *J. Schmitz, A. Merz, H. Kaspers*, Chemie-Technik **9** (1980) 229
[5] *H. Jantzen, K. Schügerl, H. Helmrich*, Chem.-Ing.-Techn. **50** (1978) 798

Isotopenpraxis **20** (1984) 2, S. 53 bis 57

The Determination of Density Variation in Radiation Protective Walls of Concrete by Means of Bremsstrahlung of 10 MeV from Linear Accelerator*)

H. Baumbach[1]), *K. Fichtler*[2]), *R. Melzer*[3]), *D. Tietze*[3]), *H. Ullrich*[4])

([1]) Leipzig University of Technology, Section Technology of Building; [2]) Academy of Building of the GDR, Institute of Structural Engineering and Foundation Engineering; [3]) Karl-Marx-University Leipzig, Radiologic Clinic and Policlinic, Physico-Technical Department; [4]) Academy of Sciences of the GDR, Central Institute of Isotope and Radiation Reserach)

Testing the function of meter-thick radiation protective walls as built e.g. for the reactor stacks of nuclear power plants shows that the relative density variations of concrete multiplied by the number of the mean free paths of the radiation in concrete have an effect on the dose rate variations behind the protective wall. It is therefore necessary to carefully control the density of concrete.
If ultrasonic testing of the walls has to be excluded because they are hermetically enclosed with a steel liner on either side and gamma-radiography fails due to the mass per unit area in the order of several tons/m², then another method, the transmission with bremsstrahlung of adequate energy, e.g. of 10 MeV, can be applied. Such bremsstrahlung can be received from a mobile and portable linear accelerator.
In the paper the first results of the determination of density variations in a model wall of concrete having masses per volume area in the range from 1800 kg · m⁻³ to 2400 kg · m⁻³ and linear dimensions 1,60 m are presented. The values have been measured in the beam direction.

Bei der Funktionstestung meterdicker Strahlenschutzwände aus Beton, wie sie z. B. beim Reaktorbau erforderlich sind, zeigt sich, daß die relative Dichteschwankung des Betons mit der Anzahl der freien Weglängen der Strahlung im Beton zum Effekt der Dosisschwankungen hinter der Strahlenschutzwand beiträgt.
Wenn Prüfungen derartiger Wände mit Ultraschallmessungen wegen eines hermetischen Abschlusses der Wand mit einem Stahlliner ausgeschlossen sind und die gammaradiographische Testung hier wegen der hohen zur durchstrahlenden Flächenmassen (einige 10³ kg · m⁻²) versagt, kann die Funktionsprüfung mittels Transmissionsmessungen von Röntgenbremsstrahlung entsprechend hoher Energien (10 MeV) erfolgen. Als Bestrahlungsquellen eignen sich mobile und transportable Linearbeschleuniger.
Es werdem erste Ergebnisse der Bestimmung von Dichteschwankungen an einer Modellwand mit Dichten im Bereich von 1800 ... 2400 kg · m⁻³ und Linearabmessungen von 1,60 m vorgestellt.

Keywords

industrial radiography; linear accelerators; concretes; containment; nuclear power plants; radiation protection

1. Introduction

The use of efficient and mobile accelerators for technical diagnostics in civil engineering will come into question wherever

1. tests cannot be carried out without radiographic and dosimetric measurements, e.g. in analyzing the state of a structure for its reconstruction, in testing special structural components in nuclear power station construction, in testing radiation protective walls, etc. and
2. the concrete thicknesses to be used (e.g. thickness >0.6 m) exclude the use of conventional X-ray and gamma radiation devices in practice owing to their too low radiant energy and dose rate.

For the use of accelerators for the transmission of objects in precasting, in processes on the building site, e.g. for the quality and function control, and in analyzing the state of structures, a number of special and general boundary conditions has to be fulfilled.

Special boundary conditions

1. A sufficient mobility of the device, i.e. especially an optimum performance-mass ratio and a compact and robust construction of the device, have to be guaranteed.
2. Connected loads (electric energy, water) have to be realizable under building site conditions.
3. The radiation protection under site conditions has to be guaranteed.

General boundary conditions

1. The relation between the expenditure for the test of structural components and the cost of the total object has to be justifiable.
2. The testing process has to be properly included in the total production-technological process from both the organizational and technological aspect.
3. The cadres required have to be trained and qualified.

---

*) Presented at the 2nd Working Meeting "Radioisotope Application and Radiation Processing in Industry", 28th September to 1st October 1982, Leipzig

Addresses: [1]) 7030 Leipzig, Karl-Liebknecht-Str. 132; [2]) 7123 Engelsdorf near Leipzig, Hans-Weigel-Str. 26; [3]) 7010 Leipzig, Liebigstr. 20a [4]) 7050 Leipzig, Permoserstr. 15

Tab. 1. Half-value thicknesses $x_{0.5}$ of concrete for varying quantum energies E, in m

E	$x_{0.5}$/m
150 keV	0.04
300 keV	0.03
1.2 MeV (Co^{60})	0.06
6 MeV	0.10
8 MeV	0.115
15 MeV	0.15
30 MeV	0.20

In Tab. 1 half-value thicknesses $x_{0.5}$ of normal concrete (mass per unit volume = 2350 kg · m^{-3}) are given for radiation energies between 150 keV and 30 MeV, which are typical of the radiography [1].

In the present paper tests for the control of the quality of structural components during the running building process are to be distinguished from function tests of the completed structure. In the case of radiation protective walls the securing of the radiation protection required supervenes on the other functions of a structure.

Several probe techniques [2] for the radiometric procedures of quality control concerning the determination of the concrete density in volume have been tested and applied. The probes in question are the plug-in probe for fresh concrete and various types of dipping probes. These techniques are primarily conceived for the quality control, and not for the dosimetric test of the biological radiation protection of parts of a building, and thus differ from irradiation procedures with bremsstrahlung, by which constructive or dosimetric statements can be obtained, in their use and field of application.

Structure parameters in different cross sections of the test object, e.g. location, diameter, and state of reinforcements in concrete, are the target quantity of radiographic investigations for the analysis of the state of a structure. From that constructive conclusions are drawn.

Relative density variations and their local distribution are the target quantity of the radiometric quality control of concrete and dosimetric function control of radiation protective structures. From that conclusions concerning the homogeneity of concrete and dose rate distributions, respectively, are drawn. In dosimetric investigations it is not, in the first place, of interest how to secure the radiation protection required, but the fact that it is really secured.

The aim of the present paper is to investigated whether and by what means it will be possible to carry out a dosimetric function control of concrete walls of more than 1 m in thickness under site conditions.

2. Foundations

The dose rate at the outlet of a linear accelerator in the forward direction depends on the particle flux density of the electron emission source and the acceleration energy. The emission of bremsstrahlung in the forward direction increases with increasing electronic energy. By electrons having an energy of 10 MeV a dose rate of about 2 · 10^3 rd · min^{-1} in a distance of 1 m from the focus can be obtained. Considering all dissipations, electrical connected loads from a few kW to some 10 kW result; these are values which are available on sites.

It should be mentioned by comparison that by a 1 Ci-Cobalt-60-source (1 Ci = 3.7 × 10^{10} Bq; mean gamma-ray energy = approx. 1.2 MeV) a dose rate of 22.5 · 10^{-3} R · min^{-1} in a distance of 1 m from the source can be obtained; this is a dose rate which is by 5 orders lower than the dose rate obtained by the above linear accelerator. Owing to the container masses required, cobalt sources of an activity < 100 Ci = 3.7 TBq are used for the mobile radiographic equipment.

The length of the accelerating tube of the linear accelerator depends on the desired electronic energy — and hence on the maximum energy of the bremsstrahlung obtained, and thus on the penetrating power of the radiation as well as on the maximum obtainable electric field strengths in the accelerator, i. e. on the electric insulation. The attainable field strengths are in the range from 10 to 20 MeV · m^{-1}. About 25 accelerating gaps of a total linear dimension below 2 m can be realized for electrons of 10 MeV. Such comparatively small dimensions are essential for the handling of these devices on building sites.

Linear accelerators do not contain any magnet for guide fields as is usual in betatrons, and, having the same dose rate, they are, therefore, lighter in weight than betatrons, which is favourable for their movement to and on the building site. Linear accelerators with an electronic energy of 10 MeV and a dose rate of 1000 rd · min^{-1} in a distance of 1 m from the focus can have masses of only 1000 kg. In a linear accelerator efficient microwave generators are used as are, for instance, employed in mobile marine stations and in the military range, which are thus robust in construction. These microwave generators determine the power consumption of the accelerator from a few kW to some 10 kW. The total power, essentially has to be dissipated as heat. For this about 10 l of water/min are required. There are also linear accelerators with a closed heat-transfer circuit, e. g. Neptun 10 MeV [3]. ЛУЭ-10 has not a closed circuit.

3. Quantitative considerations concerning the transmission of concrete members having a thickness of more than 1 m by means of bremsstrahlung

From Fig. 1 it can be seen that, in radiographic investigations of concrete (thickness of layer = 1.2 m) with bremsstrahlung of a maximum energy of 8 MeV and a dose rate of 2000 rd · min^{-1} in a distance of 1 m from the focus, an exposure time in the range of minutes has to be expected for normal macroscopic X-ray films (optical density $D = 2$). In this case, films with silver halide emulsions, scintillation arrangements, counter arrangements, luminescence processes, X-ray image converters, but also channel electron multiplier plates and silicon diode targets can be, on principle, used as radiation detectors. Films, scintillation cameras, and counter arrangements are complete and usable equipment which are available on the market at present.

As radiation detectors silver halide emulsions are inexpensive detectors (also with rising silver prices), which are flexible, especially with respect to their surface shape and form and, in addition to the radiation detection, allow the recording and documentation of the radiation fluence distributions, and make a direct visual evaluation possible. By these advantages the use of radiographic films is secured also for the future. On the other hand, processes with films (also coarse-grained and highly sensitive layers) are characterized by

1. a high local resolving power which is not fully used (by orders of magnitude) in the radiographic transmission of concrete walls of more than 1 m in thickness

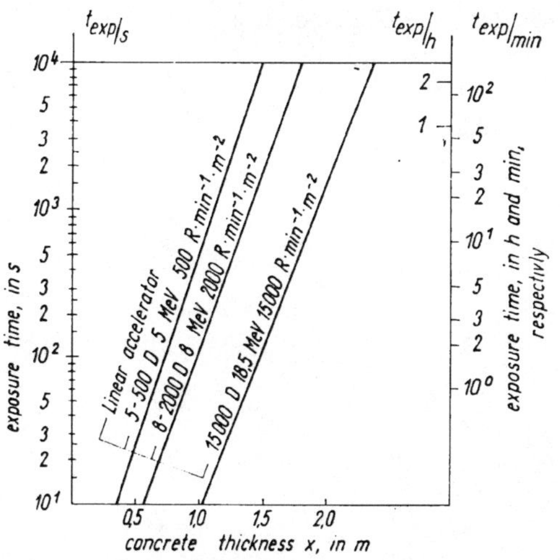

Fig. 1. Required exposure time t_{exp} in s, min, and h, respectively, for varying concrete thicknesses x in m (mass per unit volume $= 2350$ kg · m^{-3}), using linear accelerators of type ЛУЭ of NIIEFA Leningrad of 5 MeV (500 rd · min^{-1} in a distance of 1 m from the focus), of 8 MeV (2000 rd · min^{-1} in a distance of 1 m from the focus) and of 18·5 MeV (15000 rd · min^{-1} in a distance of 1 m) [4]

2. a detection probability which is lower by orders of magnitude as compared to scintillation camera or counter arrangements
3. a time lag after the results are available, and
4. the necessity of a large volume of equipment for the quantitative evaluation.

These drawbacks require a differentiated consideration of the potentials of the radiation detection for the concerns of technical diagnostics in civil engineering.

The question whether accelerators can be used also for the quality control in concreting with respect to the transmission of very thick concrete members results in two points:

1. The movements of the accelerator and detector have to correspond with the velocity of concreting.
2. The radiation protection has to be secured under these conditions.

As to the radiation detection practically only scintillation cameras or counter shields would come into question for this case; these are characterized by

1. their detection probability which is higher by 4 orders as compared to films, which leads to a reduction in the primary dose rates by this factor and thus to the range of 0·2 rd · min^{-1} in 1 m distance, which can be more easily applied:
2. the direct availability of measurement results.

X-ray films require 10^8 to 10^{10} photons/cm^2 for an avaluable optical density; by scintillation counters and counter tubes with response probabilities in the order from per cent to some 10 per cent, measured values with a standard deviation of 1% can be obtained even at 10^4 pulses recorded.

Even from estimates with greatly simplifying assumptions (narrow parallel beam) it follows that, in the case of concrete thicknesses of approximately 1 m for radiation protective walls and nuclear power station containments, changes in density multiplied by a factor of 10 are reflected in the changes of dose rate. More detailed considerations show that changes of density in a radiation protective wall have to be detected with a percentage error of a fow per cent in order to be able from that to make statements on the change of dose rate behind the wall with a percentage error of 50% what is absolutely necessary [5].

One problem of testing the function of radiation protective walls in the nuclear power station with respect to their shielding action is that there exists not any original radioactive source for the function test before the start-up of the plant, i. e. possibly over many years no statement can be made concerning the question whether the structural components erected reliably meet the function for which they were designed. For the containment original sources are available only in the case of an accident. For that reason, the only way of testing the function is to use sources which allow the detection of the dose rate to a sufficient accuracy even after 20 half-value thicknesses of the concrete (these are 1·2 metres for a gammaray emission of 1 MeV). For this purpose, practically only accelerators producing high-energy bremsstrahlung with an adequate dose rate will come into question.

Tab. 2 given a survey of performance parameters of accelerators of the following firms or institutions:

- Varian, Palo Alto, Cal., U.S.A., accelerator of type Linatron (L 200 – L 6000)
- Polytechnic Institute Tomsk, U.S.S.R., accelerator of type Betatron (П М Б 6, Б 12 – Б – 35)
- NIIEFA Leningrad, U.S.S.R., accelerator of type ЛУЭ (ЛУЭ 5 – ЛУЭ – 15)
- ZDAU IBJ Swierk, Poland, accelerator of type Neptun 10 R

Devices of Philips, Mitsubishi, and Siemens do not show any variations in the performance parameters which would be essential here.

The linear accelerator Neptun 10 of ZDAU IBJ Swierk is a production licensed by CGR-Mev, France.

For the use of a mobile accelerator for technical diagnostics in civil engineering the following performance parameters have been selected and listed here:

E — maximum energy of the accelerated electrons, in MeV
D_L — dose rate, in rd · min^{-1}, in a distance of 1 m from the focus
M — mass of the radiator head, in kg
P — electrical power required, in kW

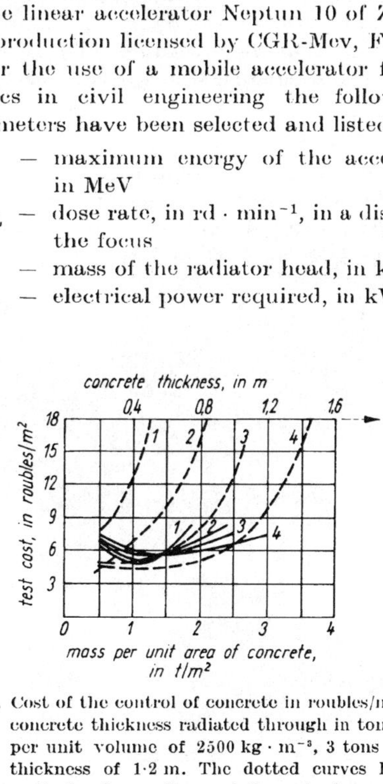

Fig. 2. Cost of the control of concrete in roubles/m^2 as a function of the concrete thickness radiated through in tons/m^2; i. e., for a mass per unit volume of 2500 kg · m^{-3}, 3 tons · m^{-2} correspond to a thickness of 1·2 m. The dotted curves hold for radiographic methods of recording and the unbroken curves, for radiometric methods of recording. The figures 1 – 4 refer to the accelerators used: 1 = ПМБ-6; 2 = Б-15; 3 = Б-25; 4 = Б-30 [7].

(It should be considered that, with, in other respects, comparable data, at least the hundredfold dose rate will be availabele if linear accelerators are used, and hence the exposure time and duration of measurement will be reduced which will have an influence on the cost, too.)

Tab. 2. Performance data of a series of accelerators and reference data to Cobalt-60 [4, 6]

Denotation	E/MeV	D_L/rd · min^{-1} (in 1 m distance from the focus)	$M_{radiator}$/kg	P/kW
^{60}Co	1·2	2·55 · 10^{-2} Ci^{-1}		
L 200	2	175	816	
L 400	4	500	900	
L 2000	8	2000	900	
L 6000	15	6000	3900	
ПМБ-6	6	0,3	100	2
Б-15	15			
Б-18, 5	18,5	10	500	5
Б-18, 5	18,5	70	2500	10
Б-25	25	1500		30
Б-35	35	1000	4500	25
Neptun 10 R	10	2000	1000	13
ЛУЭ- 5	5	500		
ЛУЭ- 8	8	2000		
ЛУЭ- 8	8	15000	2000	75
ЛУЭ-10	10	2000		
ЛУЭ-15	15	15000	5000	150
ЛУЭ-15	15	40000		

It can be seen that from the viewpoints described previously the devices of type Linatron L 200, Neptun 10 R, and ЛУЭ-8 are suitable for technical diagnostics in civil engineering.

A lead on the cost of the radiographic and radiometric control of concrete components as a function of the concrete thickness radiated through can be got from Fig. 2 [7], from wich it follows that the procedures and methods described in the paper can be applied in an economic way.

4. Experimental set-up

Before a linear accelerator of type Neptun 10 p of ZDAU IBJ with horizontally outgoing rays a concrete wall measuring 0·80 × 0·80 × 1·60 m³ in size and having masses per unit volume of 1800, 1950, 2200, and 2350 kg · m^{-3}

Fig. 3. Photo of the experimentel set-up in the radiation room of the Radiologic Clinic, Section of Medicine, Karl-Marx-University Leipzig.

On the right: Linear accelerators Neptun 10 p of ZDAU IBJ Warsaw with horizontally outgoing rays in the direction of the concrete wall
On the left: Concrete wall composed of prisms measuring 0·20 × 0·20 × 0·40 m³, having mass per unit volume of 1800, 1950, 2200, 2350 kg · m^{-3} and stairs of concrete (on the left-2nd prism from the bottom) on deals and wooden supports; total length = 1·60 m in the beam direction

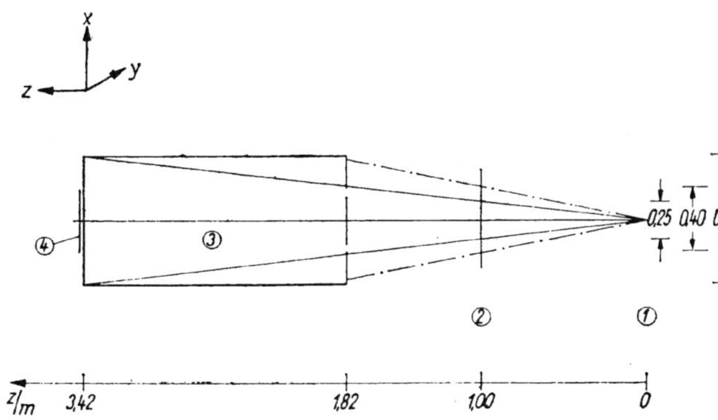

Fig. 4. Schematic representation of the experimental set-up for the irradiation of a concrete wall being 1·60 m in length with bremsstrahlung of 10 MeV

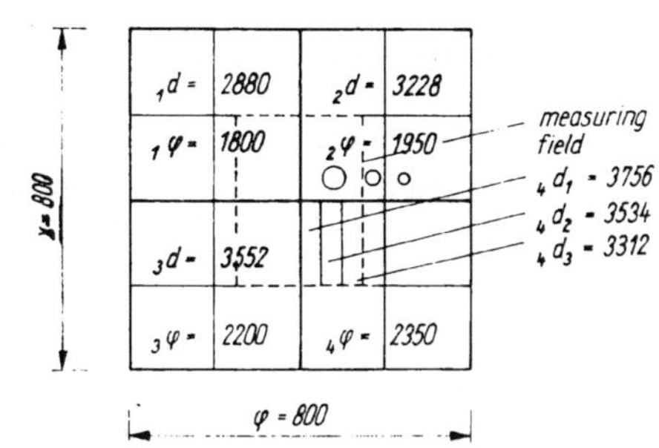

Fig. 5. Schematic representation of the arrangement and position of the concrete specimens and the measuring field

x [mm], y [mm], ϱ [kg · m^{-3}], d [kg · m^{-2}]

Bitte folgende Korrekturen beachten: Statt $_{1...4}\varphi$ lies $_{1...4}\varrho$; statt φ = 800 lies y = 800.

A photo of the experimental set-up in the radiation room of the Radiologic Clinic of the Section of Medicine, Karl-Marx-University Leipzig, is shown in Fig. 3, the schematic representation of the experimental set-up is shown in Fig. 4, and the schematic representation of the arrangement and position of the concrete specimens and measuring area in the beam direction is shown in Fig. 5.

The accelerator Neptun 10 p produces bremsstrahlung of a maximum energy of 9 MeV and a dose rate of 300 rd · min^{-1} in a distance of 1 m from the focus measured in pulses of ca. 1 rd, then the number of the impulses during the exposure time is counted, and a measure of the dose rate on the front side of the concrete block obtained. The diaphragm system at the head of the accelerator was so adjusted that at the end of the concrete block, on the side of the detector, a radiation area of $0{\cdot}80 \times 0{\cdot}80\ m^2$ resulted (Fig. 4).

ORWO X-ray film of type TF 14 having a lead foil of 0·10 mm before the film and a lead foil of 0·15 mm behind it — seen in the beam direction — was used as a radiation detector. The film has been developed according to standard requirements.

5. Test results

After an exposure time of 10 minutes corresponding to a dose of ca. 3000 rd measured in a distance of 1 m from the focus, blackenings which could be well evaluated were obtained on the ORWO X-ray film of type TF 14 in a distance of 3·42 m from the focus of the accelerator with concrete of 1·60 m being in the path of rays (comp. Figures 3—5).

The curve: film blackening as a function of the mass per unit area radiated through, in $kg \cdot m^{-2}$ has a slope of $-6{\cdot}9 \cdot 10^{-4}\ m^2 \cdot kg^{-1}$ under the above exposure conditions; i. e. under the conditions realized here (concrete thickness = 1·60 m) differences in mass per volume of 20 $kg \cdot m^{-3}$ on the detector side can be detected at best (minimum discernible blackening differences $\Delta D = 0{\cdot}02$). Owing to the X-ray scattering in concrete, this value is relevant only to density variations near the detector.

The actually obtainable values for the detection of differences in density variations in concrete do not only depend on the geometrical conditions, but also on the properties of the film (homogeneity of sensitivity), the film processing conditions (developer movement and tempering), the beam homogeneity etc. In this connection the influence of backscattering on the film is essential. Under the test conditions realized here it was of the order of the information-carrying direct radiation.

Even if only the testing technology described and conventional radiographic methods are applied, 6 m^2/h can be tested by means of the linear accelerator Neptun 10 p; i. e. in addition to the notified control of protective walls, as to the exposure time and test velocity, the method described here is also suitable for the quality control in manufacturing processes.

An essential increase in the test velocity by more than a factor of 10 and a reduction in source strengths and thus in the expenditure of the radiation protection are possible by replacing radiographic methods by radiometric procedures (Section 3).

In other words, the possible test velocity for thick-walled concrete components attained by the linear accelerator is not so much determined by the physical parameters of the source and detector, but rather by the test technology, by the hoisting and conveying technique for moving the accelerator, in particular, however, by properly including the test technology in the total building technology.

Received December 23, 1982

References

[1] *L. Runkiewicz*, Radiografia konstrukcij budowlanych z betony; Prace naukowe Instytutu Techniki Budowlanej, Seria Monografie, Rok XXXV, Wydawnictwa Instytutu Techniki Budowlanej, Warszawa 1980

[2] *E. Pohl*, Anwendung der Meßtechnik bei der Rationalisierung im KKW-Bau, Forschungsbericht 1980; IfI 230/77-80; THL-STB 02-01-5

[3] Persönliche Mitteilung von *A. Ciszek*, Direktor des ZDAU IBJ Swierk (3. 12. 1980)

[4] 1.) Доклады: Второго Всесоюзного Совещания по применению ускорителей заряженных частиц в народном хозяйстве, Ленинград 1—3 окт. 1975 г.; Государственный комитет по использованию атомной энергии СССР; Научно-исследовательский институт электрофизической аппаратуры имени Д. В. Ефремова. Ленинград 1976

2.) Доклады: Третьего Всесоюзного совещания по применению ускорителей заряженных частиц в народном хозяйстве, Ленинград 26—28 июня 1979 г.: Государственный комитет по использованию атомной энергии СССР; Научно исследовательский институт электрофизической аппаратуры имени Д. В. Ефремова. Ленинград 1979

[5] *W. Krüger*, SAAS Berlin, 2. Leipziger Bauphysikalisches Seminar 1981, interne Mitteilung

[6] Бариман, Радиационный отдел, Пало Алто, Калифорния, США: Использование пентгеновского излучения высоких энергии в установках Линатрон

[7] *В. А. Воробъев*, Применение физико-математических методов в исследовании свойства бетона; Москва, „Высшая школа" 1977

Isotopenpraxis 20 (1984) 2, S. 58 bis 60

Metallbestimmung in ^{14}C-markierten zinnorganischen Verbindungen durch Neutronenaktivierung*)

D. Klötzer, W. Görner

(Zentralinstitut für Kernforschung Rossendorf der AdW der DDR)[1]

Beschrieben wird die Neutronenaktivierungsanalyse (NAA) als geeignete Methode zur Zinnbestimmung. Das Anwendungsbeispiel umfaßt Photolyseexperimente am System [1-^{14}C]Tributylzinnoxid/Cellulose, wobei die NAA in Kombination mit anderen Analyseprinzipien sowohl bei kinetischen Untersuchungen zur Photolyse dieses Systems als auch bei der Strukturaufklärung von Zwischenprodukten wesentliche Beiträge liefert.

This paper deals with the non-destructive short-time neutron activation analysis of tin. It includes experiments of photolysis on the system [1-^{14}C]tributyltinoxide/cellulose, in which the neutron activation combined with other analytical principles backs kinetic investigations of system photolysis and helps to clear up the structure of intermediate products.

Keywords

analysis; neutron activation; tin; organic tin compounds; carbon 14; photolysis; ultraviolet radiation

Anwendungstechnische Untersuchungen mit Organozinnverbindungen, z.B. die Aufklärung der Lichtstabilität des Systems Tributylzinnoxid/Cellulose, werfen einige analytische Probleme auf, da die klassischen Methoden im praxisbezogenen Konzentrationsbereich der Organozinnwirkstoffe kaum brauchbar sind. Aus diesem Grund wurden viele der in den letzten Jahren zu dieser Problematik veröffentlichten Arbeiten — z.B. zur Photolyse des Systems Tributylzinnoxid/Cellulose — mit Hilfe der ^{14}C-markierten Organozinnverbindung durchgeführt [1—4]. Durch kombinierten Einsatz von Radiodünnschichtchromatographie, Radiogaschromatographie, verschiedener Methoden der Radioaktivitätsmessung sowie der Neutronenaktivierungsanalyse konnten eine Reihe von wichtigen Ergebnissen zur Kinetik und zum Mechanismus der Photolyse dieses Systems gewonnen werden.

Um den Fortgang der Photolyse dieses Systems zu studieren und entsprechende Zwischenprodukte zu charakterisieren, wurde die Neutronenaktivierungsanalyse als Methode zur Zinnbestimmung modifiziert.

Experimentelles

Das im natürlichen Zinn mit einer Häufigkeit von 5,8% vorkommende Isotop $^{124}_{50}Sn$ wird im Core des Rossendorfer Kernreaktors bei einem thermischen Fluß von ca. $6 \cdot 10^{13}$ n $\cdot$ cm^{-2} $\cdot$ s^{-1} durch Neutroneneinfang in das kurzlebige Isotop $^{125m}_{50}Sn$ umgewandelt. Die Bestrahlungszeit im Reaktorcore beträgt etwa 1 min. Benutzt wird folgende Reaktion:

$$^{124}_{50}Sn(n, \gamma)^{125m}_{50}Sn$$

Die Halbwertszeit von $^{125m}_{50}Sn$ beträgt 9,5 min [5]. Nach einer Abklingzeit von 5—30 min wird im Gammaspektrometer die 332 keV-γ-Linie selektiv gemessen und mit einem Standard verglichen. Als Standards wurden Tributylzinnoxid/Cellulose- und $SnCl_2$/Cellulose-Proben mit Zinngehalten und 100 von 10000 ppm präpariert. Die Analysenproben werden in Polyethylenbeutel (max. 50 mm × 50 mm) eingeschweißt bzw. bei Flüssigkeitsproben mit Hilfe von Polyethylenschläuchen für den Bestrahlungs- u. Meßvorgang konfektioniert.

Tab. 1. Meßreihen an Analysenproben mit bekanntem Zinngehalt

Probe	vorgegebener Zinngehalt ppm	Einwaage mg	Analysenergebnis ppm
Tributylzinnoxid/Cellulose	10000	204,0	8587
	5000	573,9	3694
	100	1809,5	112
	50	3747,0	47,1
	10	5122,0	15,9
Tributylzinnbenzoat/Cellulose	10000	231,0	8310
	100	1993,0	88,6
$SnCl_2$/Cellulose	10000	437,5	7218
	1000	1152,5	840
	100	2270,0	88,4
	50	3667,0	51,9
	10	5130,0	36,2

Zur Überprüfung der Methode wurden Meßreihen an Analysenproben mit bekanntem Zinngehalt durchgeführt (Tab. 1). Die Proben wurden so dimensioniert, daß der Zinngehalt mindestens 0,2 mg betrug. Die in Tab. 1 aufgezeichneten Ergebnisse belegen die Brauchbarkeit der neutronenaktivierungsanalytischen Zinnbestimmung. Die untere Nachweisgrenze lag bei 10 ppm. Die mittlere Abweichung des gemessenen vom vorgegebenen Zinngehalt ist hauptsächlich auf die Methode der Celluloseimprägnierung mit organischen und anorganischen Zinnverbindungen zurückzuführen, wie statistisch durch Vermessung mehrerer Proben der gleichen Mischung ermittelt wurde (Tab. 2).

*) Vortrag auf der 3. Tagung „Nukleare Analysenverfahren" vom 11. — 15. 4. 1983 in Dresden

[1]) Anschrift: 8051 Dresden, PF 19

Tab. 2. Streuung der Analysenergebnisse verschiedener Proben der gleichen Mischung

Mischung ppm Sn	Probe	Meßwert ppm	Mittelwert $\bar{x}$ ppm	relat. Standard-abweichung	Zählstatist. Streuung der NAA	Abweichung des $\bar{x}$ v. Sollwert
	1	9980				
1000	2	8310	9197	9,2%	0,6%	−8,0%
	3	9300				
	1	82,0				
100	2	88,6	87,3	5,4%	3,6%	−12,7%
	3	91,4				

Durch Kombination dieser Methode mit anderen Analysenprinzipien konnten von der gleichen ^{14}C-markierten Bestrahlunsprobe bzw. vom gleichen Experiment mehrere wichtige Informationen über den Abbaumechanismus und die -geschwindigkeit gewonnen werden. Radiogaschromatographisch [2] konnte beispielsweise nachgewiesen werden, daß sich die photolytisch von [1−^{14}C]Tributylzinnoxid abgespalteten [1−^{14}C]Butylradikale vorzugsweise zu n-[1−^{14}C]Butan stabilisieren. Verwendet wurde dazu ein Gaschromatograph Giede 10,3 WLD; komplettiert durch Verbrennungsofen (Kobaltoxid, 650 °C), H_2O-Adsorptionsrohr, Radiodetector [2]. Mit Hilfe der Radiodünnschichtchromatographie [1] konnte der photolytische Abbau des Biocids [1−^{14}C]Tributylzinnoxid über die [1−^{14}C]Dibutylzinn- bis zur [1−^{14}C]Monobutylzinnstufe verfolgt werden. Der letzte Abbauschritt zur Stufe der anorganischen Zinnverbindungen wurde neutronenaktivierungsanalytisch (nach vorheriger chromatographischer Abtrennung der Organozinnkomponenten) beobachtet.

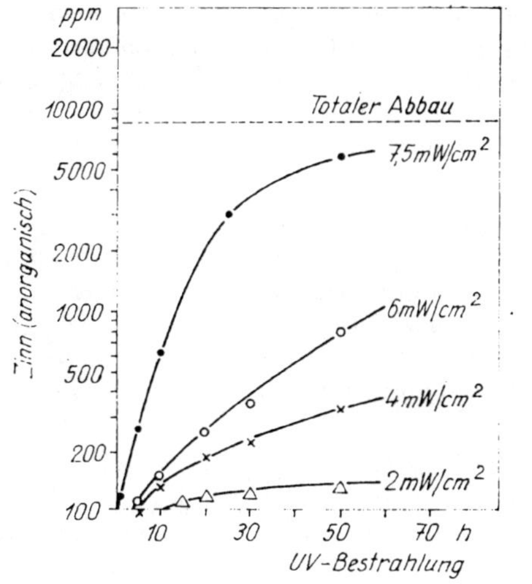

Abb. 1. Abhängigkeit der Photolysegeschwindigkeit von der Bestrahlungsintensität

Versuchsbedingungen: Xenonhöchstdrucklampe, $\lambda > 280$ nm, Luftatmosphäre, 45 °C

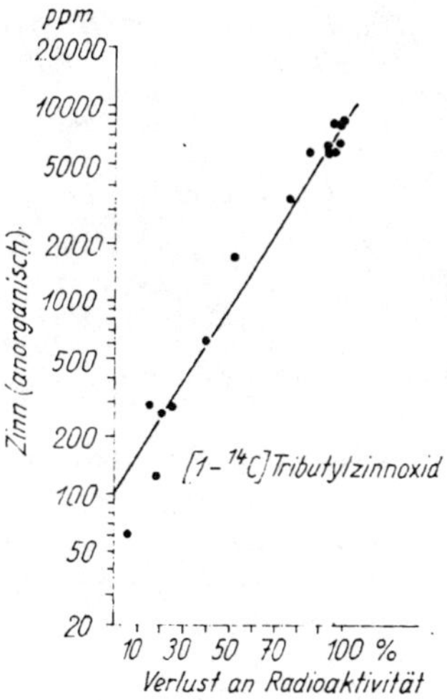

Abb. 2. Korrelation von Radioaktivitätsverlust und Zunahme des Gehaltes an anorganischem Zinn bei UV-Bestrahlung des Systems [1−^{14}C]Tributylzinnoxid/Cellulose

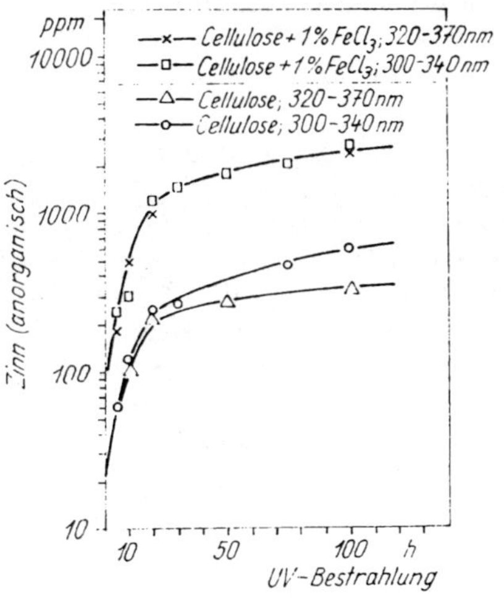

Abb. 3. Photolysegeschwindigkeit des Systems [1−^{14}C]Tributylzinnoxid/Cellulose in Abhängigkeit vom Reinheitsgrad der Cellulose (eisenhaltig und eisenfrei)

Versuchsbedingungen: 6 mW/cm², Luftatmosphäre, 45 °C

Wie Abb. 1 zeigt, ist die Photolysegeschwindigkeit für [1−^{14}C]Tributylzinnoxid im System [1−^{14}C]Tributylzinnoxid/Cellulose bei konstantem Wellenlängenbereich des eingestrahlten Lichtes beträchtlich von der Bestrahlungsintensität abhängig. Bemerkenswert ist, daß dabei die Radioaktivitätsabnahme und die Zunahme des Gehaltes an anorganischem Zinn nahezu linear korrelieren (Abb. 2). Somit beschreibt jeder dieser Parameter für sich allein den photochemischen Gesamtprozeß. Auf ähnliche Weise wurde der Einfluß des Reinheitsgrades der Cellulose auf die Photolysegeschwindigkeit des Systems untersucht. Hier wurde kommerzielle Cellulose MN 300 (Firma Macherey u. Nagel, BRD) mit 1% $FeCl_3$ imprägniert [3]. Die Bestrahlung des Systems [1−^{14}C]Tributylzinnoxid/Cellulose erfolgte hier mit langwelligerem UV-Licht ($\lambda > 300$ nm) als bei den bisherigen Photolyseexperimenten.

Wie aus Abb. 3 ersichtlich ist, wurden die eisenhaltigen Systeme mit größerer Photolysegeschwindigkeit abgebaut als die eisenfreien Systeme.

Gute Dienste leistete die neutronenaktivierungsanalytische Zinnbestimmung auch bei der Strukturaufklärung von Zwischenprodukten der Photolyse des Systems [$1-{}^{14}C$] Tributylzinnoxid/Glucose [4]. So konnte ein solches photochemisches Reaktionsprodukt mit Hilfe von IR-Spektroskopie, Radiodünnschichtchromatographie, neutronenaktivierungsanalytischer Zinnbestimmung und elementaranalytischer Kohlenstoffbestimmung als Bis{[$1-{}^{14}C$]Tributylstannyl} glucoseether identifiziert werden. Der gemessene Zinngehalt lag bei 30,6% (theoret.: 31,4%).

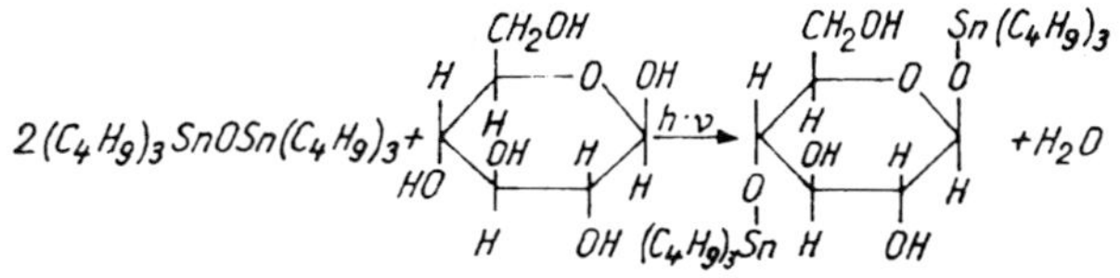

Zusammenfassung

Die Neutronenaktivierungsanalyse als zerstörungsfreie Methode liefert in Kombination mit anderen Analyseprinzipien wesentliche Beiträge zum photochemischen Abbaumechanismus des Systems Tributylzinnoxid/Cellulose sowie zur quantitativen Erfassung des Einflusses einzelner Parameter auf die Photolysegeschwindigkeit von Tributylzinnoxid. Die gute Eignung der Neutronenaktivierungsanalyse zur Bestimmung metallischer Heteroatome (z.B. Zinn) eröffnet Möglichkeiten der Struktur-, Mechanismus- und Kinetikaufklärung, deren Bearbeitung nicht unbedingt an den Standort eines Kernreaktors gebunden ist.

Eingegangen am 22. 6. 1983

Literatur

[1] *D. Klötzer*, Isotopenpraxis **12** (1976) 128
[2] *D. Klötzer, H. Görner, K.-H. Heise*, Isotopenpraxis **15** (1979) 134
[3] *D. Klötzer*, Isotopenpraxis **16** (1980) 56
[4] *D. Klötzer*, Isotopenpraxis **18** (1982) 102
[5] *W. Kunz, J. Schintlmeister*, „Tabellen der Atomkerne" Teil I, Bd. I, Berlin: Akademieverlag 1958

Isotopenpraxis **20** (1984) 2, S. 60 bis 64

Untergrundlaboratorium für Messungen geringer Radioaktivität*)

W. Helbig, S. Niese, D. Birnstein

(Zentralinstitut für Kernforschung der AdW der DDR, Rossendorf)[1])

Unter 47 m Gesteinsbedeckung entsprechend 125 metres water equivalent — daraus resultiert eine mehr als 40fache Schwächung der Höhenstrahlungsintensität — wurde ein massives Labor auf der Fläche 6 × 3,5 m² mit 3,5 m Höhe errichtet. Auf einem 0,8 m hohen Serpentinitsockel wurden Wände und Dachbelegung von je 0,7 m Dicke als Stahlkonstruktion mit Füllung von Serpentinit ausgeführt. Eine Bleiburg von 10 cm Wandstärke umschließt einen Ge(Li)-Detektor von 75 cm³ und 9,1% Effektivität. Das Nulleffektspektrum ist mit Literaturwerten von Antikoinzidenzanordnungen vergleichbar. Nachweisgrenzen für 12 Stunden Meßzeit liegen für kleine Proben, die den Untergrund im Meßbereich nicht erhöhen, bei 0,025 Bq; für die natürliche Radioaktivität in 1000 g-Gesteinsproben bei 0,15 ppm Th, 0,03 ppm U und 0,011% K_2O.

An underground laboratory has been installed under 47 m rock covering (125 metres water equivalent), where the cosmic radiation ist diminished more than 40-fold. The massive laboratory building, 6 × 3·5 m² by 3·4 m height, was erected on a foundation of 0·8 m serpentinite with a framework of old steel, forming walls and ceiling with 0·7 m thickness, filled with serpentinite. A Ge(Li)-detector of 75 m³ and 9·1% efficiency ist shielded by lead of 10 cm thickness. The background spectra are comparable with those of anticoincidence devices described in the literature. Detection limits for small samples amount to 0·025 Bq; for natural radioactivities of 1000 g samples to 0·15 ppm Th, 0·03 ppm U and 0·011% K_2O for 12 h-measurements.

Keywords

background radiation, cosmic radiation, gamma spectroscopy, laboratory equipment, low level counting, natural radioactivity, underground

1. Einleitung

Die Verminderung des Nulleffektes kernphysikalischer Detektoren durch Absorption des Höhenstrahlungsanteils in massiven Abschirmungen von Untergrundlabors ist bereits in mehreren Ländern genutzt worden [1—7]. In einigen dieser Labors wurde durch sehr große Gesteinsbedekkung (660 ... 5000 metres water equivalent, mwe) die Höhenstrahlung praktisch ausgeschaltet [1, 2, 3b], in anderen durch Schichten von 60 .... 100 mwe eine teilweise Schwächung erreicht [3—5, 7]. So wird bei 70 mwe von 10facher [3] und 14facher [4] Verminderung berichtet. Aus einigen Veröffentlichungen geht nicht hervor, ob stationäre Labors eingerichtet wurden [5, 6].

In allen diesen Labors waren zusätzliche aufwendige Abschirmungen notwendig, um durch Absorption der Strahlung der natürlichen Radioaktivität der Umgebung eine effektive Reduzierung des gesamten Nulleffektes zu bewirken. Zur Raumauskleidung wurden ultrabasische Gesteine wie Dunit [1] und Serpentit [3], die sehr niedrige Gehalte an natürlichen radioaktiven Nukliden habe, von 0,5 bis 1 m Dicke verwendet, teilweise mit ausgesuchtem Zement verarbeitet; Blei von 10 cm Stärke [2] oder Stahl von 20 cm Dicke [4]. Aufwendige Abschirmungen aus Blei und anderen üblichen Materialien umgeben die Detektoren.

*) Vortrag auf der 3. Tagung „Nukleare Analysenverfahren" vom 11.—15. 4. 1983 in Dresden

[1]) Anschrift: 8051 Dresden, PF 19

Mitteilungen / Information

BERICHT UEBER DIE TAETIGKEIT DER IAEA VOR DER UNO IN NEW YORK

Am 4. November 1983 erstattete der Generaldirektor der IAEA Dr. H.Blix, vor der Generalversammlung der Vereinten Nationen in New York Bericht.
Er ging u.a. auf folgende Aspekte naeher ein:

Kernenergiegewinnung:
Zwar haetten einige Staaten dadurch, dass die Wirtschaftskrise allmaehlich staerker fuehlbar werde als die Energiekrise, ihre Kernenergieprogramme eingeschraenkt, trotzdem betonte Dr. Blix: "Wenn die wirtschaftliche Erholung weitergehen und die Entwicklung an Boden gewinnen soll, werden wir mehr Energie brauchen, und die Welt muss weiter trachten, sich von ihrer uebermaessigen Abhaengigkeit vom Oel zu befreien. Dazu bieten heute nur die Kohle und die Kernenergie Moeglichkeiten von einiger Bedeutung."Am besten kann die IAEA die Kernenergiegewinnung foerdern, wenn sie mithilft, ihre Wirtschaftlichkeit,Sicherheit u. Umweltfreundlichkeit noch zu verbessern, sagte Dr. Blix. "Waehrend des Jahres 1982 ist es, wie in anderen Industrien,durch Ausfall von Geraeten oder menschliches Versagen zu potentiell gefaehrlichen Betriebsstoerungen gekommen, aber sie konnten ausnahmslos durch die Sicherheitssysteme unwirksam gemacht werden."

Entsorgung:
Auf der Konferenz in Seattle(Mai 1983) sei man unter anderem zu der Erkenntnis gelangt, dass man der Entsorgung bei Abfaellen geringer und mittlerer Aktivitaet weiterhin Aufmerksamkeit schenken muss. Eine Frage habe sich im Zusammenhang mit der Endlagerung hochaktiver Abfaelle ergeben: "Es kann Laender geben, die keine geologisch geeigneten Standorte fuer die Endlagerung hochaktiver Abfaelle besitzen,oder deren Kernenergieprogramm zu unbedeutend ist, um die Errichtung eines Endlagers wirtschaftlich tragbar zu machen ... Sollen sie deshalb auf die Kernenergie verzichten oder waere das Problem durch internationale Zusammenarbeit zu loesen? Vielleicht wird es bald an der Zeit sein, sich die Moeglichkeiten fuer eine internationale Zusammenarbeit bei der Endlagerung hochaktiver Abfaelle zu ueberlegen."

Technische Zusammenarbeit:
Dr. Blix erlaeuterte ein 1982 angelaufenes Projekt zur Ausrottung der Mittelmeerfruchtfliege in Aegypten - das groesste Vorhaben, dessen Durchfuehrung die IAEA bisher uebernommen hat. Durch dieses Programm,bei dem Bestrahlungsverfahren angewandt werden und das einen Aufwand von mehr als 20 Mill. US-Dollar auf vier Jahre vorsieht, sollen die bedeutenden Verluste vermieden werden, die jedes Jahr in der aegyptischen Obst- und Gemueseproduktion eintreten.

Sicherheitskontrolle:
Die Bestrebungen der IAEA, durch Sicherheitskontrolle die Weiterverbreitung von Kernwaffen zu verhindern, sind etwas Einzigdastehendes - nicht nur im Rahmen ihrer Taetigkeit, sondern auch als Beispiel eines internationalen "Aufbaues von Vertrauen", meint Dr. Blix. "Man muss sich darueber im klaren sein, dass die Sicherheitskontrolle der IAEA, so wichtig sie auch fuer den Aufbau des Vertrauens sein mag, inhaerenten Beschraenkungen unterliegt: sie gibt auch keine vollstaendige Gewaehr, dass in Laendern, wo nur einige Anlagen ueberwacht werden, keine Materialabzweigung erfolgt. Trotzdem erlaubt sie es, bei jeder Art von Aktivitaeten zu kontrollieren, ob sie friedlich ist, und Alarm zu schlagen, falls es Anzeichen fuer eine Materialabzweigung geben sollte. Die wesentlichste Verbesserung der Sicherheitskontrolle der IAEA waere eine Erweiterung ihres Wirkungsbereiches."

(IAEA-Pressekommunique (1983) 27)

DUBNA-REAKTOR BIETET NEUE FORSCHUNGSMOEGLICHKEITEN

Am Vereinigten Kernforschungsinstitut in Dubna ist ein neuer Reaktor "IBR-2"in Betrieb genommen worden. Unter den modernen Kernforschungsreaktoren der verschiedensten Typen nimmt der "IBR-2" einen besonderen Platz ein:er ist eine prinzipiell neue Konstruktion und liefert gepulste Neutronenstrahlen; er arbeitet z.Z. bei 1000 kW, ist aber fuer wesentlich hoehere Leistungen ausgelegt. In den Intervallen zwischen den Impulsen betraegt seine Kapazitaet etwa 60 000 kW, waehrend sie waehrend der Impulse, von denen jeder 2.10^{-4}s dauert, 700 000 kW erreicht. Der Impulsstrom ist damit waehrend des Strahlungsmaximums fuenfmal staerker als die Leistung der bisher besten stationaeren Reaktoren in der Welt. Sehr umfangreich ist das geplante Programm der Erforschung der Struktur und Atombewegung von festen und fluessigen Koerpern.

(Sowjetische Pressemitteilung (1983))

UdSSR: 1500-MWe-BLOCK KRITISCH

Am 4.10.1983 wurde eine 1500 MW_e-Einheit Ignalinsk-1 in der Lettischen SSR zum ersten Mal kritisch. Dies ist die bisher groesste Reaktorblock-Einheit der Welt. Der Reaktor selbst ist ein graphitmoderierter Siedewasserreaktor mit Druckkanaelen, die in der Sowjetunion entwickelt wurden.

(Presse der Sowjetunion, Okt. 1983)

REGIONALE ZUSAMMENARBEIT IN LATEINAMERIKA

in fuenf lateinamerikanischen Mitgliedsstaaten der IAEA - Bolivien, Ekuador, Kolumbien, Peru und Venezuela - werden mit den zustaendigen Behoerden Vereinbarungen ueber die Erstellung eines koordinierten regionalen Programms getroffen, das die Weitergabe kernwissenschaftlicher und -technischer Errungenschaften an diese Laender foerdern soll. Mit Hilfe der IAEA soll eine enge Zusammenarbeit der Laender in Form gemeinsamer Forschungsprojekte zustandekommen, und zwar auf den Gebieten Strahlenschutz, kerntechnische Geraete, Radioisotopentechnik, Verwendung von Forschungsreaktoren und Bestrahlungstechnik.
Alle fuenf Laender befinden sich kerntechnisch annaehernd auf derselben Entwicklungsstufe. Sieben Reaktoren oder Teilchenbeschleuniger fuer hohe Energiebereiche sind in diesen Laendern in Betrieb oder im Bau, und in allen sind Arbeiten ueber Verwendung von Isotopen und Strahlen auf verschiedenen Gebieten im Gange, die fuer die Gesundheit oder den wirtschaftlichen Wohlstand der Bevoelkerung von Bedeutung sind.(IAEA-Pressekommunique (1983)25)

WISSENSCHAFTLICHE TAGE 1984 DER TECHNISCHEN HOCHSCHULE "OTTO VON GUERICKE" MAGDEBURG

IV. Symposium "ZUVERLAESSIGKEIT"
am 4. und 5. September 1984

Themenkomplexe:
- Zuverlaessigkeit von Erzeugnissen,
- Technische Diagnostik und Schadensbekaempfung

Tagung "MATHEMATIK FUER DIE PRAXIS"
vom 11. bis 14. September 1984

Themenkomplexe:
- Erhoehung der Breitenwirksamkeit mathematischer Verfahren und interaktiver Arbeitstechniken,
- Vervollkommnung und Weiterentwicklung mathematischer Verfahren und interaktiver Arbeitstechniken, die sich in der Praxis bewaehrt haben,
- Erschliessung neuer Einsatzgebiete fuer die Mathematik und fuer die interaktiven Arbeitstechniken.

(Auskuenfte erteilt: TH "OTTO VON GUERICKE" Magdeburg, Tagungsbuero, 3040 Magdeburg, PSF 124)

Kurznachrichten / Newlets

Oekologische Mikrobiologie:
INTENSITY OF MICROBIOLOGICAL CH_4 oxidation in NATURAL ECOSYSTEMS

Fractionation of carbon isotopes of CH_4 as the result of its microbiological oxidation in a flowing gas feeding system was determined. It was shown that during CH_4 consumption by microorganisms its residual part was enriched in C-13 relatively to its content in the initial CH_4 was proportional to the percentage of its consumption by microorganisms. Proportionality coefficient was 25 +- 5‰. The change in the isotopic composition of carbon of CH_4 can be also used for the quantitative determination of its share oxidized by methanotrophic bacterias in natural ecosystems.
(A.M. Zyakun et al., Geokhimia (1983) 5 and Dokl. Akad. Nauk 272 (1983) 5)

Trennung der Wasserstoffisotope:
RESEARCH ON THE SEPARATION OF H ISOTOPES BY CATALYTIC EXCHANGE

Separation of hydrogen isotopes is essential to the design and operation of CANDU nuclear reactors fueled with natural uranium. Wetproofed platinum catalysts, which retain their performance in the presence of liquid water have proven effective in promoting isotope exchange between countercurrent streams of hydrogen and liquid water. They offer considerable scope for development of new processes to extract deuterium from natural water and to remove tritium from both light and heavy water. Isotopic exchange in packed catalyst columns has been modeled in terms of the rates of isotope transfer between hydrogen and water vapour and between water vapour and liquid. Measurements of kinetic equilibrium isotope effects arising from these studies are discussed.
(J.H. Rolston, in: Synthesis and Applications of Isotopically Labeled Compounds, Proceedings of an International Symposium, Kansas City, MO, USA, 6-11 June 1982, W.P. Duncan and A.B. Susan (Eds.), 385-390, 1983 Elsevier Scientific Publishing Company, Amsterdam)

CF_3H/HDO-Isotopenaustausch:
CATALYSIS OF DEUTERIUM EXCHANGE BETWEEN FLUOROFORM (CF_3H,D) AND WATER (THE FEED STEP FOR A POTENTIAL DEUTERIUM SEPARATION PROCESS)

Several classes of reagents have been examined for their ability to catalyze deuterium exchange between gaseous fluoroform and water. Included were NaOH (alone, with dimethyl sulfoxide (DMSO) as co-solvent, and with tetraalkylammonium salts as phase transfer agents), anionic ion exchange resins, and some types of supported metal catalysts (e.g. Pt, Pd on carbon or alumina) known to be effective for H_2/HDO exchange. The metal catalysts were found to be unreactive for CF_3H/D_2O exchange, but all of the alkaline systems showed at least some reactivity above 50°C. The use of aqueous DMSO solvents yields reasonably fast deuterium transfer rates.
(E.A. Symons et al., in: Synthesis and Applications of Isotopically Labeled Compounds, Proceedings of an International Symposium, Kansas City, MO, USA, 6-11 June 1982, W.P. Duncan and A.B. Susan (Eds.), 411-412, 1983 Elsevier Scientific Publishing Co., Amsterdam)

Radiooekologie:
BRD: RADIOOEKOLOGISCHE UNTERSUCHUNGEN AN PFLANZEN

Im Kernforschungszentrum Karlsruhe, BRD, wurde ein Gewaechshaus mit 150m² Nutzflaeche (Kosten 1,5 Mill. DM) in Betrieb genommen, in dem der Transfer radioaktiver Stoffe aus dem Boden in Pflanzen erforscht werden soll. Als Nutzpflanzen werden Weizen, Gerste, Roggen, Hafer, Mais, Futterrueben, Kartoffeln, Weisskohl, Gruenkohl, Blumenkohl, Kopfsalat, Spinat, Moehren, Zwiebeln, Gurken, Tomaten, Erbsen, Bohnen und Erdbeeren untersucht, fuer deren Aufzucht bis zu acht verschiedene, mit Radionukliden verunreinigte Bodensorten verwendet werden. Ziel dieser Versuche ist es, die Belastung der Bevoelkerung durch radioaktive Stoffe in der Nahrung abschaetzen zu koennen und Standardwerte zu erarbeiten, die eine zuverlaessigere Berechnung der Strahlenbelastung als bisher erlauben. Zunaechst soll untersucht werden, wie sich Bodenart, chemische Bindung der Radionuklide und in den Boden eingebrachte Komplexbildner auf die Verfuegbarkeit der radioaktiven Substanzen fuer die Pflanzen auswirken. Weiterhin ist vorgesehen, den Einfluss von Duengemitteln, Wachstumsdauer und Bodennutzung zu untersuchen und die raeumliche Verteilung der Bodennuklide in der Pflanze zu bestimmen.
(Biowiss. Inform. 7 (1983) 9 - nach Frankfurter Allg. Ztg. (BRD) v. 13.7.1983)

Fallout:
Eu-155 AND Y-88 IN FALLOUT SAMPLES

The dry deposition samples as well as the precipitation ones are a rich source of information about the atmospheric pollutants. By gamma ray spectrometry it is possible to reach very low detection limits for these samples. During the period January-December 1981 a daily dry deposition and precipitation (if the case) sample was collected using a collector with an area of 2,7m². Eu-155 may occur as daughter of Sm-155 and directly from fission as well: Sm-155(22,1 min)-->Eu-155(4,96y)--> Gd-155(stable). The Eu-155 activity in sample corresponds to a fallout of about 50 Microcurie . km-2 . year^{-1}.
The activity of Y-88 in the sample was (15+-8) pCi, which corresponds to (5,5+-2,8) pCi.m^{-2}. The presence of these radionuclides in the atmosphere is correlated which the thermonuclear test that took place in Oct. 1980. It is to be noted that the measurement of a similar sample collected during 1980 gives no evidence about the presence of Eu-155 and Y-88. (C. Dovlete et al.; Atomkernenergie/Kerntechnik 43 (1983) 3)

Praeparative Chemie:
HERSTELLUNG VON ETHAN-[1,1-D_2] AUS 1,1-DILITHIOETAN

Nahezu qualitativ reines Ethan-[1,1-D_2] laesst sich aus 1,1- Bis(chloromercurio)ethan herstellen, das mit Lithiumpulver in wasserfreiem Diethylether bei Raumtemperatur umgesetzt und anschliessend mit D_2O zersetzt wird. Die Herstellung von 1,1-Bis(chloromercurio)ethan kann auf verschiedene Weise erfolgen; am einfachsten aus 1,1-Dichlorethan. (A. Maercker et al., Angew. Chemie 95 (1983) 9)

Veröffentlichungen / Publications

STAATLICHES AMT FUER ATOMSICHERHEIT UND STRAHLENSCHUTZ DER DDR:

SAAS - 302
Dosimetrie inkorporierter Transurane (Dosimetry of incorporated transuranic radionuclides) - 289 S. - V. Loessner

SAAS - 307
Accounting for and control of nuclear material at the Central Institute of Nuclear Research-Rossendorf - 17 S. - S. Heidel et al.

SAAS - 308
Verfahren zur Bearbeitung radioaktiver Sonderabfaelle (Methods for the treatment of radioactive special wastes) - 18 S. -
W. Koerner und Chr. Herzog

Veranstaltungskalender – Tagungen – Lehrgänge / Calendar – Meetings – Courses

RADIATION PROCESSING FOR PLASTICS & RUBBER II
International Conference at the University of Kent at Canterbury, UK, 28. March to 30. March 1984
Sessions: A-Scientific, B-Processing, C-Applications (Grafting), D-Applications (Medical), E-Applications (General), Poster Papers

Ausgewählte Beiträge / Recently published papers

ATOMNAJA ENERGIJA 56(1984)1

System Synthesis to Stabilize Energy Distribution and Control Reactor Power on the Basis of Side Ionization Chambers (Russ.) - I.Ya. Emelyanov et al.

Analysis of Statistical Error and Optimization of Correlation Flowmeters (Russ.) - B.V. Kebadze

Analytical Approximation of Neutron Physics Data (Russ.) - S.A. Badikov et al.

The Hydrogen Balance in Reactor INTOR (Russ.) - V.M. Sharapov et al.

The Application of Quartz Glass for Ionizing Radiation Dosimetry (Russ.) - R.R. Gulamova et al.

Application Package for Data Analysis of Environment Protection and Human Health (Russ.) - E.I. Vorobyov et al.

Effects of Radionuclide Deflation on Contaminated Areas for One-Time and Steady Releases Into the Atmosphere (Russ.) - K.P. Makhonko

The Application of Proton and Deutron Activation Methods of Analysis for the Determination of Elements with Z>42 (Russ.) - S. Mukhammedov et al.

On the Transfer of Radioactive Substances Along Roads by Transportation Wheels (Russ.) - V.E. Vostrukhov et al.

ATOMNAJA ENERGIJA 55(1983)5

Measurement of the Total Neutron Cross-Section and Resonance Parameters of Cf-249 in the 0,02-66 eV Energy Range(Russ.) - V.A.Anufriyev et al.

High-Voltage Electron Accelerator "Aurora IV" (Russ.) - B.I. Albertinsky et al.

Electron Accelerators with Local Shielding for Radiation Technology (Russ.) - B.I.Albertinsky et al.

A Method of Calibration in Radio-activation Analysis of the Elemental Composition of a Substance (Russ.) - Vo Dak Bang

Quantitative Determination of the Limit of Nuclide Observation by Instrumental Neutron Activation Analysis of Complex Samples with Epithermal Neutrons(Russ.) - V.I.Drynkin et al.

The Effect of Selective Excitation in Photon Activation Analysis Using X-ray Spectrometry (Russ.) - V.G. Magera and M.G. Davydov

Methods for Predicting Remote Biological Effects and Estimation of Genetic Consequences of Radiation Exposure (Russ.) - Yu.I.Moskalyov and V.F. Zhuravlyov

The Comparison of Methods for Determining the Concentration of Rn-222 Decay Products in Air (Russ.) - M.V. Terentyev and E.M. Krisyuk

Radiation Chemical Stability of CsI (Russ.) - I.A. Kulikov and M.L. Malyshev

Comparative Radiation Characteristics of Natural and Recycled Thorium(Russ.) - L.V.Matveyev et al.

Measurement of the B-10 Content in Nuclear Reactor Absorption Elements by the Registration of Prompt Gamma Quanta of Li-7 (Russ.) - V.P. Korolyova and P.S. Otstavnov

ATOM (1983)323

Technology transfer - the importance of partnership - R.G. Sowden

WISSENSCHAFT UND FORTSCHRITT 33(1983)9

Biologisch-medizinische Strahlenforschung (Zum 100. Geburtstag von Walter Friedrich) - H. Abel

Es gibt kein "Quantenraetsel" der Photosynthese (Zum 100. Geburtstag von Otto Warburg) - E. Hoextermann

Eichbosonen experimentell nachgewiesen
G. und J. Ranft

Zukunftsaufgaben der Hydrogeologie -
E.V. Pinneker

J. CHEM. PHYSICS 79(1983)7

The infrared multiphoton excitation and photochemistry of DN_3 - T.B. Simpson et al.

A systematic study of the convergence and additivity of correlation and basis set effects on the force constants of small molecules: HF, HCN and NH_3 - P. Pulay et al.

The vibrational spectra of solid II ethane and ethane-d_6 - M.G. Wisnosky et al.

Tracer diffusion of counterions through a solution of polyelectrolyte
K. Zero et al.

J. CHEM. PHYSICS 79(1983)8

C-13 spin diffusion of adamantan
Ch.E. Bronniman et al.

Reduced dimensionality quantum rate constants for the $D + H_2$ ($\nu = 0$) and $D + H_2$ ($\nu = 1$) reactions on the LSTH surface - J.M. Bowman et al.

Mobilities of Na^+ ions in H_2-D_2 mixtures at room temperature - K. Jinuma et al.

Deuterium NMR of biaxial discotic liquid crystals - D. Goldfarb et al.

THEOCHEM 13(1983)9

Compliant fields for water, ammonia & methane - A.S.N. Murthy and S. Ranganathan

The formate anion: an ab initio study of its equilibrium structure, general harmonic force field and fundamental vibrational frequencies - A.G. Gregory et al.

GEOLOGY 11(1983)9

Origin and age of Adirondack anorthosites - re-evaluated with Nd isotopes
A.R. Basu and H.S. Petingill

Radiometric dating of ash partings in coal of the Eocene Puget Group, Washington: Implications for paleobotanical stages
D.L. Turner et al.

NEW ZEALAND J. GEOLOGY & GEOPHYSICS 26(1983)3

A He isotope profile of Lake Taupo, New Zealand - T. Torgersen

Age of the volcanoes & granite basement of the Auckland Islands, Southwest Pacific -
C.J. Adams

The taxonomic status, and isotopic evidence for paleoenvironments, of giant oysters from the Oligocene Te Kuiti Group, South Auckland, New Zealand - C.S. Nelson et al.

ECONOMIC GEOLOGY 78(1983)6

Pb-isotope systematics of strata-bound sulfide deposits in the Higher Nappe Complexes of the Swedish Caledonides
K. Sundblad and M.B. Stephens

Sr-isotope evidence for the origin of barites & sulfides from the Mississippi Valley - Type ore deposits in Southeast Missouri
S. Lange et al.

GEOPHYSICAL RESEARCH LETTERS 10(1983)10

Identification of D ions in the ionosphere of Venus - R.E. Hartle and H.A. Taylor

PALEOGEOGRAPHY, PALEOCLIMATOLOGY, PALEOECOLOGY 43(1983)1/2

Isotopic paleohydrology of Wadi el Akarit (Southern Tunisia) during the late Pleistocene and Halocene - J.Ch. Fontes et al.

J. GEOPHYSICAL RES. 88(1983)C13

Use of long lived Ru daughters as indicators of exchange between the free troposphere and the marine boundary layer - M.A. Kritz

Temporal distributions of radiostrontium isotopes and Rn daughters in rain water during a thunderstorm -
L.A. Burchfieldt et al.

Rn-222 from the Ocean surface-C. Duenas et al.

CONTRIBUTIONS MINERALOGY PETROLOGY 83(1983)3/4

Sr-87/Sr-86 in Kimberlitic carbonates by ion microprobe: hydrothermal alteration, crustal contamination and relation to carbonalite -
R.A. Exley and A.P. Jones

The oxygen isotope composition of Hercynian granites and pre-Hercynian gneisses from Schwarzwald, SW Germany -
J.Hoefs and R.Emermann

Rb-Sr-dating of alteration events in granitoids - F.E. Wickmann et al.

J. PHYSICAL OCEANOGRAPHY 13(1983)7

A tritium box model of the North Atlantic thermocline - J.L.Sarmiento

RADIOCARBON 25(1983)1

GEOSECS Indian Ocean and Mediterranean Radiocarbon - M. Stuiver and H.G. Oestlund

DEUTSCHE MEDIZINISCHE WOCHENSCHRIFT 108(1983)39

NMR-Spektroskopie in der Medizin (eine Uebersicht mit 24 Zit.) - J. Frahm u.a.

GASTROENTEROLOGY 84(1983)6

Autoradiographic demonstrations of estrogen receptors in the esophagus of baboons -
W.B. Winborn et al.

PROC. INDIAN NATIONAL SCIENCE ACADEMY 49(1983)A2

Vibrational analysis of nitryl halogenides -
S. Mohan and K.G. Ravi Kumar

PRECAMBRIAN RES. 21(1983)1/2

The geochemical and oxygen-isotope affinities of Proterozoic mafic granulites from the Einasleigh metamorphics, northern Queensland -
N.J. McNaughton and A.F. Wilson

J.GEOPHYSIC. RES. 88(1983)B 9

Age and significance of Precambrian basement samples from northern Illinois and adjacent states - W.J. Hoppe et al.

Oxygen isotope study of Precambrian granites from the Illinois-Deep-Hole-Project. -
Y.N. Shieh

Oxygen isotopic compositions of minerals from Illinois Deep Drill Hole granites -
M. Kombrink and R.N. Clayton

Isotopic composition of U & Th in crystalline rocks - J.N. Rosholt

GEOCHIM. COSMOCHIM. ACTA 47(1983)8

Laboratory studies of the diagenesis and mobility of Pu-239, Pu-240 and Cs-137 in nearshore sediments - E.R. Sholkovitz et al.

Pb isotope geochemistry of a massif-type anorthositic-Charnockitic body -the Hidra Massif (Rogaland, SW Norway) - D.Weis and D. Demaiffe

Effects of thermal maturation on stable organic carbon isotopes as determined by hydrous pyrolysis of Woodford Skale - M.D. Lewan

The calcite-graphite isotope thermometer - Data on graphite bearing marbles from Naxos,Greece-
R. Kreulen and P.C.J.M. Vanbeek

The use of O-18/O-16 ratios to study the formation and chemical origin of coal -
J.Dunbar and A.T. Wilson

GEOCHIM. COSMOCHIM. ACTA 47(1983)9

Stepwise fluorination - a useful approach for the isotopic analysis of hydrous minerals -
M. Haimsan and L.P. Knauth

CHEMICAL GEOLOGY 40(1983)3/4

A nuclear method for the analysis of light elements ($Z \leq 11$) in geological samples -
M. Basutcu et al.

GEOPHYSICAL RES.LETTERS 10(1983)9

Oxygen and silicon isotopes in ALHA 81005 -
T.K. Mayeda et al.

Recent cosmic ray exposure history of ALHA 81005 (the 1st Lunar meteorite) - C. Tunitz et al.

COMPUTERS AND GEOSCIENCES 9(1983)3

Numerical modelling of trace element fractionation during diffusion controlled crystallization - C.J. Wright et al.

Für die Einrichtung eines Untergrundlabors wurde ein ungenutzter Stollen ausgewählt, der mit ca. 6 m Breite und ebensolcher Höhe 60 m in ein steil ansteigendes Gesteinsmassiv hineinführt. Über dem für den Laborbau ausgesuchten, relativ trockenen Stollenteil, 50 m vom Eingang entfernt, beträgt die Gesteinsbedeckung 47 m entsprechend 125 mwe (Abb. 1).

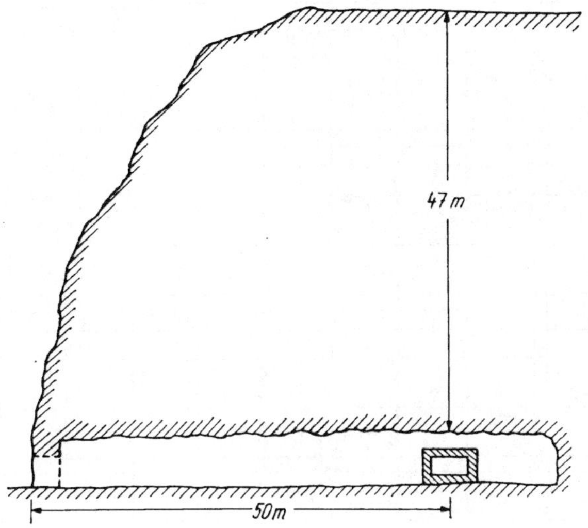

Abb. 1. Geländeprofil mit Untergrundlabor

Aus Angaben über die Schwächung der Höhenstrahlung mit der Tiefe kann unter diesen Bedingungen der Faktor 40 erwartet werden. Zur Schwächung der natürlichen Radioaktivität des Gesteins sollte der Faktor 10^4 für die 2614 keV-Strahlung des ^{208}Tl angestrebt werden. Aus der Strahlenschutzliteratur kann entnommen werden, daß eine 0,7 m dicke Betonwand mit einer zusätzlichen Bleiabschirmung von 10 cm Stärke diese Forderung etwa erfüllt. Tritt Serpentinit — mit etwa 2 Größenordnungen geringeren Gehalten an U, Th und K, verglichen mit Beton oder Oberflächengesteinen wie Granit — an Stelle des Betons, so ist zu erwarten, daß der verbleibende Nulleffekt im wesentlichen durch radioaktive Verunreinigungen im Detektor und seiner Umgebung verursacht wird. Die Abschirmung gegen die Umgebungsstrahlung wäre dann ausreichend.

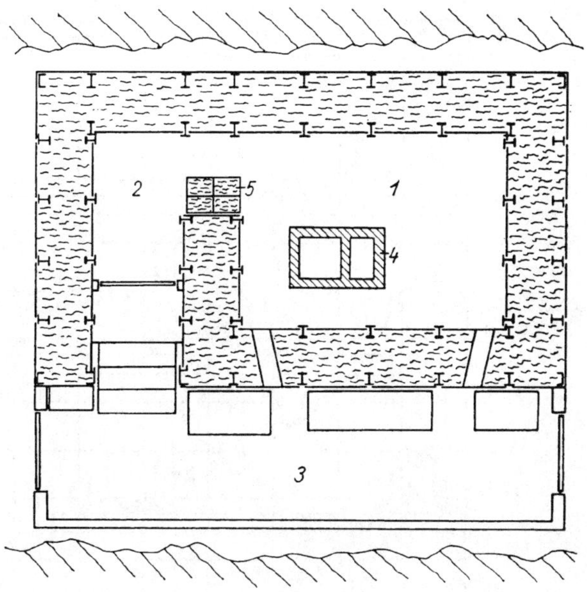

Abb. 2. Schematischer Grundriß des Labors

1 — Innenraum; 2 — Schleuse; 3 — Meßraum; 4 — Bleiburg; 5 — Serpentinitquader

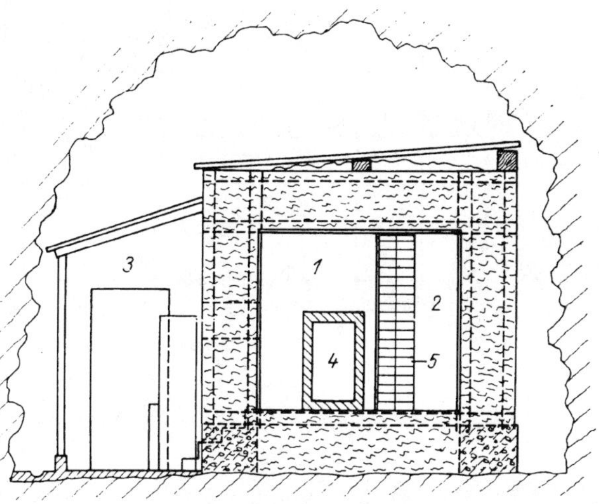

Abb. 3. Schnitt durch das Labor

2. Beschreibung

Die Größe des Labors wurde nach den erwähnten Abschätzungen mit Rücksicht auf die vorhandenen Raumverhältnisse und die zu installierenden Meßeinrichtungen festgelegt (Abb. 2, 3). Auf dem Boden des Stollens wurde ein Sockel von 0,8 m Höhe und 0,6 m Breite errichtet, der eine Fläche von 6 × 3,6 m² umfaßt. Als Zuschlagsstoff diente Serpentinitsplitt verschiedener Korngrößen. Zement PZ 7/375 wurde wegen seiner relativ niedrigsten Konzentration natürlicher Radioaktivität unter verschiedenen Sorten ausgewählt. Auf dem Sockel wurde ein Stahlgerüst vorwiegend aus alten Eisenbahnschienen aufgebaut. Es besteht aus der Sockelbelegung, doppelten Wänden mit der Gesamthöhe 3,4 m und der Dachbelegung. Die Innenflächen wurden mit 14 mm starken Altstahlplatten aus Abriß-Eisenbahnbrücken belegt, die Außenflächen mit 6 mm dickem Blech aus neuem Stahl. Die so entstandenen Wände von 0,7 m Dicke wurden, ebenso wie der Sockelinnenraum, mit Serpentinitbruchsteinen und -splitt verschiedener Korngrößen verfüllt, wobei eine möglichst große Verdichtung durch Rammen mit Holzstempel und Hammer vorgenommen wurde. Die Dachbelegung wurde bis zu einer Höhe von rd. 0,85 m lose aufgefüllt. Die Gesamtmasse des Serpentinits (rd. 100 t ohne Sockel) und das verfüllte Volumen lassen den Rückschluß auf eine mittlere Dichte von 2,3 t/m³ der verdichteten Bereiche zu.

Der Innenraum hat eine Grundfläche von 2,2 × 3,1 m² und 2 m lichte Höhe, er ist durch eine Eingangsschleuse von 0,95 m Breite zugängig. Diese ist am inneren Teil durch Serpentinitquader auf 0,5 m verengt. Angrenzend an eine Langseite des Baus wurde ein Meßraum von 6 m Länge × 1,5 m Breite in Leichtbauweise errichtet. Er nimmt in Regalen die Elektronik sowie andere Geräte und Material auf. Drei rechteckförmige Durchbrüche von 0,2 m Breite und 0,4 m Höhe sind in die Wand zwischen Innenraum und Meßraum eingearbeitet, sie sollen später den Einbau einer Probenwechselautomatik ermöglichen. Zur Zeit sind sie mit Serpentinitquadern zugestellt. Weitere Durchgangsöffnungen sind für Kabel und evtl. Be- und Entlüftung eingebaut. Türen an den Schmalseiten des Meßraumes bilden die Verbindung zu den angrenzenden Stollenbereichen. Der gesamte Bau wurde mit Wellasbestplatten nach oben

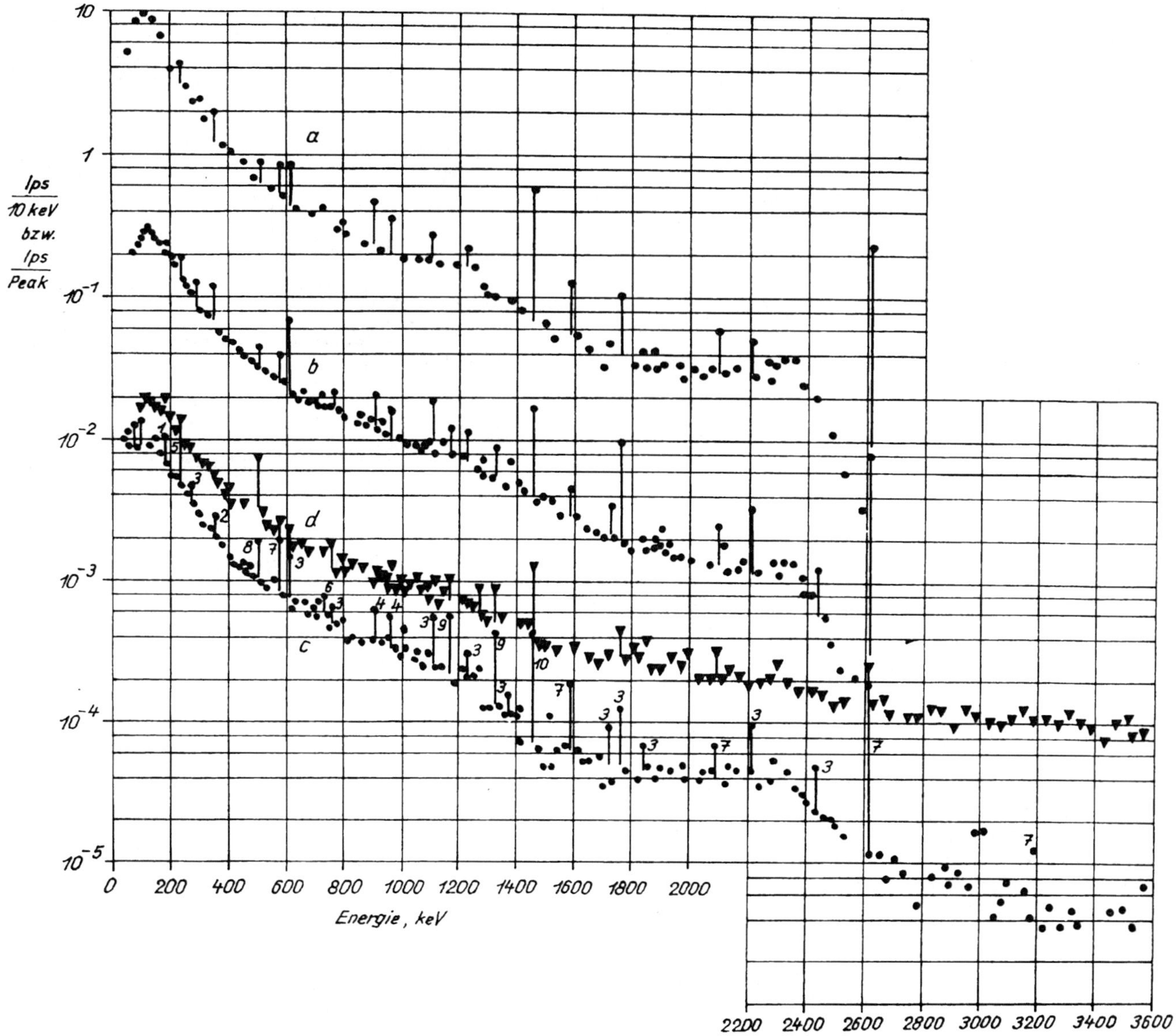

Abb. 4. NE-Spektren des 75 cm³ Ge(Li)-Detektors im Untergrundlabor

a) außerhalb des Laborbaus im Stollen
b) innerhalb des Laborbaus, ohne Bleiabschirmung
c) in der Bleiabschirmung des Labors (Meßzeit: 10^6 s)

Zum Vergleich wurde in einem Zf K-Kellerraum der NE mit der gleichen Bleiabschirmung gemessen (d).
Der Ordinatenmaßstab Ips/10 keV erlaubt die Peakflächen unabhängig von der Detektorauflösung direkt abzulesen [18].
Im NE-Spektrum c) sind enthalten

Linien der ^{238}U-Reihe:
1 – ^{226}Ra (zugleich ^{235}U) (186); 2 – ^{214}Pb (352); 3 – ^{214}Bi (275, 609, 768, 1120, 1238, 1378, 1728, 1764, 1848, 2204, 2440)
Linien der ^{232}Th-Reihe:
4 – ^{238}Ac (911, 969); 5 – ^{212}Pb (239); 6 – ^{212}Bi (727); 7 – ^{208}Tl (583, 1592 (DE), 2103 (SE), 2614, 3197 (2614 + 583))
ferner:
8 – der Annihil.-Peak (511); 9 – ^{60}Co (1173, 1332); 10 – ^{40}K (1460)
Die in Klammen gesetzten Zahlen geben die Energien in keV an.

abgedeckt zum Schutz gegen evtl. einfallendes Tropfwasser. Durch den Betrieb der z. Zt. installierten Meßgeräte und einer Zusatzheizung werden Temperatur (20 °C) und relative Feuchtigkeit (rd. 70%) konstant gehalten.

Im Innenraum wurde eine zweiteilige Abschirmkammer aus Bleisteinen aufgebaut. Der größere Teil mit 0,7 × 0,7 m² Grundfläche und 1,1 m Höhe umgibt einen Ge(Li)-Detektor von 75 cm³ samt Dewargefäßt allseitig mit 10 cm starken Wänden. Innerhalb dieser Kammer sind einige weitere 5 cm dicke Bleisteine aufgestellt, ferner umgibt den Detektor eine Haube aus 2 cm dickem Elektrolytkupfer. An diese Abschirmung wurde eine Nebenkammer aus gleichartigen Wänden mit der Grundfläche 0,7 × 0,4 m² und 0,7 m Höhe angesetzt, die Szintillationsdetektoren aufnehmen kann. Sie wurde auf einen Sockel aus Serpentinitquadern von 0,4 m Höhe aufgesetzt, damit sie die gleiche Höhe wie die Hauptkammer erreicht. Die obere gemeinsame Abdeckung, wiederum von 10 cm Stärke, ist in der Mittelachse von einer Rollbahn unterbrochen, auf der einzelne Bleiteile von je 20 kg leicht über beide Kammern wie auch auf Ablaufplatten bewegt werden können, die wahlweise an beiden Seiten der Kammer ansetzbar sind. Durch die Rollbahn kann ein Spalt von 15 cm Breite freigelegt werden, der das Einsetzen von Proben und Eingriffe in den Bereich der Detektoren erlaubt. Die Längsachse der Kammern und der Rollbahn liegt parallel zur großen Laborwand. Die Versorgung des Ge(Li)-Detektors mit flüssigem Stickstoff erfolgt durch eine Rohrleitung von außerhalb der Bleikammer.

3. Nulleffektmessungen und Nachweisgrenzen

Nulleffekt (NE)-Spektren wurden gemessen im nicht abgeschirmten Stollen (Abb. 4a), im Laborinnenraum ohne Bleikammer (Abb. 4b) und innerhalb der Bleikammer (Abb. 4c).

Das Spektrum c muß für Abschätzungen der Leistungsfähigkeit herangezogen werden. Es zeigt gegenüber dem einer konventionellen Meßeinrichtung im ZfK-Labor [8] im Bereich bis 2,5 MeV eine etwa gleichbleibende Verminderung um den Faktor 25.

Da die Bleiabschirmung im Niederniveaulabor jedoch wesentlich aufwendiger ist als die im ZfK-Labor, wurde sie zum Vergleich vorübergehend in einem Kellerraum des ZfK aufgebaut und das NE-Spektrum gemessen (Abb. 4d).

Während die Verminderung der Höhenstrahlungsintensität insgesamt im Untergrundlabor als Verhältnis der integralen Zählraten oberhalb 10 MeV mit einem NaI(Tl)-Detektor zu 43 bestimmt worden war [9], beträgt die Verringerung des NE im Vergleich der Spektren c) und d)

bei 3 bis 4 MeV	rd. 20 fach
bei 1,4 bis 2 MeV	rd. 6fach
bei 1,1 bis 1,2 MeV	rd. 3,5fach
bei 0,3 bis 1 MeV	rd. 3fach.

Die nach niedrigeren Energien scheinbar geringer werdende Abschirmwirkung im Untergrundlabor muß den vorwiegend natürlichen radioaktiven Verunreinigungen im Detektor selbst, in seiner Nähe und im Abschirmmaterial zugeschrieben werden. Untersuchungen hierüber sind mit dem Ziel der weiteren Reduzierung des NE vorgesehen. Integrale Werte des NE-Spektrums wurden jeweils bis 4 MeV summiert (Tab. 1).

Tab. 1. Integrale Werte des Nulleffektspektrums, jeweils bis 4 MeV summiert, für Ge(Li)-Detektor 75 cm^3 im Untergrundlabor

Schwelle MeV	Nulleffekt-Rate Imp/h
3	1,8
2	11
1,5	20
1	55
0,5	180
0,2	520

Das im Kellerraum des ZfK gemessene Spektrum d) ist mit entsprechenden Daten anderer Autoren vergleichbar, die effektive Abschirmungen anwendeten. So ist das von *Hunt* [10] abgebildete NE-Spektrum — ohne Antikoinzidenz — mit einem Ge(Li)-Detektor gleicher Größe um Faktoren 1,6 (bei 600 keV) bis 2,0 (bei 1500 keV) höher. Nahezu identische Verhältnisfaktoren ergeben sich nach *Camp* [11] für einen Low-level-Meßplatz, während für eine Ultralow-Einrichtung der gleichen Arbeit unter 1 MeV Faktoren bei 1 gelten, bei 2 MeV rd. 1,4. Der Detektor hatte 15% Effektivität. Von *Cooper* [12] für einen Detektor von 14,3% Effektivität gemessene Werte liegen bei 600 keV 2fach, bei 1500 keV 2,6fach höher als die entsprechenden im Spektrum d).

NE-Vergleichsspektren von Ge(Li)-Detektoren in anderen Untergrundlabors sind uns nicht bekannt. Das Untergrundspektrum c) mit NE-Spektren von Antikoinzidenzeinrichtungen zu vergleichen ist nur bedingt gerechtfertigt, da deren Effektivitäten vom Zerfallschema des Meßnuklids abhängen. In [10] und [12] ablesbare Werte solcher Einrichtungen liegen — etwa übereinstimmend — bei 600 keV bei 0,5, zwischen 1000 und 1500 keV bei 1 und um 2000 keV bei 1,4, verglichen mit dem Spektrum c). Auch weitere Daten aus [14—16] lassen die annähernde Gleichwertigkeit — hinsichtlich des erreichbaren Nulleffektniveaus — der Abschirmung in einem Untergrundlabor und durch Antikoinzidenzeinrichtungen erkennen, die schon von *Alexeev* u. a. [17] aus NaI(Tl)-Messungen geschlossen wurde. Außergewöhnliche Abschirm- und Antikoinzidenzaufbauten verbunden mit der Auswahl besonders aktivitätsarmer Materialien auch für die Detektorkonstruktion [16] bringen eine NE-Verminderung um Faktoren 0,5 bis 0,8, verglichen mit dem Spektrum Abb. 4 c), jedoch für einen Detektor von 25% Effektivität.

Die Vergleiche lassen den Schluß zu, daß eine weitere Verminderung des NE im Untergrundlabor durch Verwendung ausgesuchter Materialien noch möglich ist, aber im wesentlichen im Energiebereich <1500 keV und in begrenztem Maß.

Tab. 2a. Nachweisgrenzen im Untergrundlabor für Nuklide, deren γ-Energien nicht mit NE-Linien zusammenfallen, bei 100% Übergangswahrscheinlichkeit
Probe direkt auf dem Detektor, Meßzeit 12 h

Energie keV	NWG Bq
300	0,015
500	0,016
800	0,023
1000	0,025
1200	0,027
1500	0,022
2000	0,025

Tab. 2b. Nachweisgrenzen für Th, U und K in Proben verschiedener Massen für 12 h Meßzeit bei Auswertung einzelner Linien

	Energie keV	Probenmasse 20 g	66 g	1000 g
Th ppm	239	0,9	0,34	0,08
	583	2,8	0,9	0,14
	911	2,9	1,0	0,17
	969	4,4	1,5	0,19
	2614	2,2	1,0	0,15
U ppm	295	0,9	0,35	0,06
	352	0,5	0,2	0,03
	609	0,5	0,2	0,03
K_2O %	1460	0,15	0,064	0,011

Aus dem NE-Spektrum c) lassen sich Nachweisgrenzen als Aktivitäten ableiten. Voraussetzung dafür ist, daß im Bereich der zu messenden Linien der Untergrund durch die Probe nicht nennenswert erhöht wird. Nach *Currie* [13] wurde das detection limit $L_D^* = 2{,}71 + 3{,}29\sqrt{\mu_B}$ berechnet, wobei für μ_B der NE für 5 keV (als Peakbreite) eingesetzt wurde. Die erhaltenen Werte wurden durch die Detektoreffektivität (Probe direkt auf dem Detektor) geteilt (Tab. 2a).

Beispiele für die Nachweisgrenzen der natürlichen radioaktiven Elemente U, Th und K lassen sich aus Messungen des Standardgesteins GnA ableiten. Hierzu wurden für μ_B die entsprechenden Brutto-Intensitäten einzelner Linien des NE-Spektrum eingesetzt (Tab. 2b). Als Meßzeiten wurden in beiden Fällen 12 Stunden angenommen.

Die 20 g- und 66 g-Proben des GnA waren in Plastbechern von 50 mm Durchmesser direkt auf den Detektor gestellt worden. Die 1000 g-Probe war in einem Spezialgefäß in Bohrlochform eingesetzt.

Für den Aufbau des Labors sind wir besonders den Herren Dipl.-Ing. *Timme*, Dipl.-Ing. *G. Fischer* und Ing. *B. Schröder* zu Dank verpflichtet.

Eingegangen am 2. 3. 1983
Überarbeitete Fassung am 25. 5. 1983

Literatur

[1] *E. L. Kovalchuk, V. V. Kuzminov, A. A. Pomansky, G. A. Zatsepin*, Proc. Int. Conf. Low-Radioactivity Meas. and Appl. High Tatras 1975, Proceedings Bratislava 1977, S. 23

[2] *W. Kolb*, wie [1], S. 219

[3a] *H. Oeschger, J. Beer, H. H. Loosli, U. Schotterer*, IAEA-SM-252/13

[3b] *H. Oeschger, H. H. Loosli*, wie [1], S. 13

[4a] *S. Tanaka, W. Sakamoto, M. Takagi*, Nucl. Instr. Meth. **56** (1967) 319

[4b] *K. Yamakoshi, K. Nogami*, Nucl. Instr. Meth. **134** (1976) 519

[5] *I. H. Kaye, F. P. Brauer, R. E. Connally, H. G. Rieck*, Nucl. Instr. Meth. **100** (1972) 333

[6] *M. Alessio, L. Allegri, F. Bella, S. Improta*, Nucl. Instr. Meth. **137** (1976) 537

[7] *J. C. Barton*, 14. Int. Cosmic Ray Conf., Vol. 6, AED Conf. 75-696-000 (1975) 2155

[8] *W. Helbig, W. Görner, S. Niese*, 2. Int. Conf. Low Level Counting, High Tatras 1980

[9] *W. Helbig, W. Görner, H. Kleeberg, S. Niese*, 2. Tagung Nukl. Analysenverfahren Dresden 1979, Preprints 183

[10] *G. J. Hunt, M. C. O'Riordan, P. D. Whetmath*, Nucl. Instr. Meth. **156** (1978) 573

[11] *D. C. Camp, C. Gastrousis, L. A. Maynard*, Nucl. Instr. Meth. **117** (1974) 189

[12] *J. A. Cooper, R. W. Perkins*, Nucl. Instr. Meth. **99** (1972) 125

[13] *L. A. Currie*, Anal. Chem. **40** (1968) 586

[14] *G. Heusser, W. Hampel, M. Hübner*, 2. Int. Conf. Low-level Counting High Talras 1980, Proceedings Vol. 1, 127

[15] *J. Adam, B. Kracik, A. Kugler*, 2. Int. Conf. Low-level Counting High Talras 1980, Proceedings Vol. 1, 127

[16] *N. A. Wogman*, IAEA-Sm-252/25, (1981) S. 15

[17] *V. A. Alexeev, V. D. Gorin, T. Dimchev, A. L. Irlier, I. V. Smirnov N. G. Triumfor, O. S. Tsvetkov*, 2. Int. Conf. Low-level Counting High Talras 1980, Proceedings Vol. 2, 71

[18] *W. Görner*, J. Radioanal. Chem. **58** (1980) 61

Isotopenpraxis **20** (1984) 2, S. 64 bis 68

Isotopic Composition of Groundwaters from Kufra (Libya) as Indicator for Groundwater Formation

F. M. Swailem[1]), *M. S. Hamza*[2]), *A. I. M. Aly*[2])

(Middle Eastern Regional Radioisotope Centre, Cairo)

The results of the isotopic composition of shallow and deep groundwaters from Kufra region indicate the fossil origin of these waters and that they are not recharged under the present climatic conditions. The virtual absence of tritium and the radiocarbon ages of these waters show that they were formed mainly in the past pluvial periods. Deuterium and oxygen-18 results indicate that the groundwaters were recharged under cooler climatic conditions. These results may explain the origin of the large amounts of groundwater which existed in the region.

Die Untersuchung der Isotopenzusammensetzung flacher und tiefer Grundgewässer in der Region Kufra ergab den fossilen Ursprung dieser Gewässer sowie die Tatsache, daß sie sich nicht unter den gegenwärtigen klimatischen Bedingungen aufgefüllt haben. Das praktisch gänzliche Fehlen von Tritium und das Radiokarbonalter dieser Gewässer zeigen, daß sie sich im wesentlichen nach der Pluvialzeit gebildet haben. Die Ergebnisse der Deuterium- und ^{18}O-Ermittlung weisen darauf hin, daß sich die Grundwässer unter kühleren klimatischen Bedingungen aufgefüllt haben. Diese Erkenntnisse könnten den Ursprung der großen Mengen Grundwasser, die in dieser Region vorhanden sind, erklären.

Keywords

aquifers; carbon 14; deuterium; ground water; hydrology; isotope dating; isotope ratio; Libya; oxygen 18; tritium

Introduction

Kufra area lies in the southeastern part of Libya at an altitude of about 415 m above sea level (Fig. 1). The area occupies a gentle northeast trending depression of about 50 km long and 30 km wide. Kufra oasis is centrally located in the Kufra basin, which extends 400 km to the borders of Sudan and Chad and nearly 200 km to the borders of Egypt. The prevailing climate is extremely arid. Huge quantities of water were discovered in the oasis in 1968 by Occendental Oil company while drilling for oil. The Libyan government subsequently planned an agro-industrial project for developing the area; this project comprises drilling 200 wells to produce 200 million gallons of water per day.

[1]) Department of Physics, Faculty of Appl. Sciences and Eng., Makkah, P.O. Box 3711, Saudi Arabia.

[2]) Atomic Energy Authority, Nuclear Chemistry Dept., Cairo, Egypt.

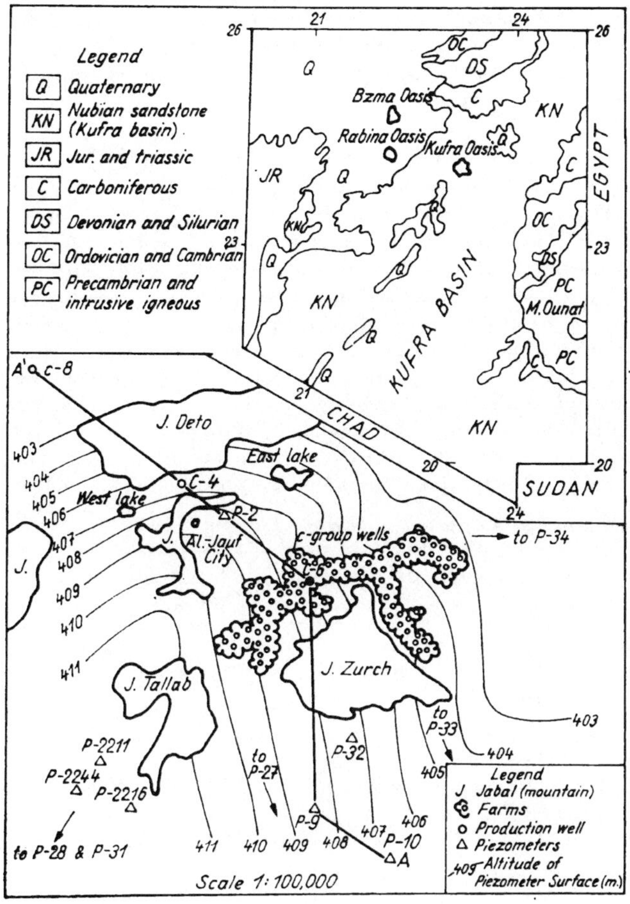

Fig. 1. Location map of Kufra and the sampling sites

The capacity of the groundwater reservoir of the Kufra basin has been estimated at 3×10^{12} cubic meters [1]. The quality of this water is suitable for domestic use. The origin and recharge processes of these huge amounts of potable water have to be investigated. Hydrological studies indicated that the piezometric configuration of the aquifer shows a small slope towards the northeast [2], (Fig. 2). Moreover, tracer experiments, using single borehole technique, carried out in some localities in the region reported a low natural groundwater movement in the order of 1.7 m per year towards the northeast [3]. Several authors estimated different recharge rates for a long scale groundwater movement from the rainy mountainous terrian at Tibisti and Aintibi in the south [1–6].

Geological and hydrological situation

Kufra basin is the easternmost Paleozoic-Mesozoic rock basin of the Sahara. It covers a large part of the southeastern Cyrenaic and extends southward into Chad and southeastward into Sudan. The rock exposed in Kufra basin range in age from Precambrian to Cretaceous (Fig. 1) [2, 4, 5].

The Nubian sandstone is the main aquifer in Kufra/basin and covered by recent sand deposits. Its thickness ranges from 850 to 1700 m and is composed of sequences of unconsolidated sandstones, siltstones and claystones of continental origin and of Mesozoic age [4]. The Nubian sandstone overlies the upper Paleozoic sediments which are about 1300 m thick.

According to the hydrogeological section (Fig. 2), the Nubian sandstone quifer in Kufra basin is classified into three general units [5]: a lower deep aquifer, an intervening aquiclude or confining layer; and a shallow water table aquifer.

The upper aquifer is a water table shallow aquifer and occurs at depths between 60 to 80 m below ground level. It varies in depth from well to well and is less permeable than the deep aquifer due to the existence of large amounts of silt and clay.

The lower aquifer is a deep semi-confined aquifer having a depth between 122 to 290 m below ground level. It consists of more permeable sandstone and conglomerate beds and is under low artesian pressure. The grain sizes, generally, range from fine to coarse and a minor amount of silt and clay is present.

Between the upper and the lower aquifers, the middle part of the Nubian sandstone is a zone of low permeability consists of silt and caly sediments. It acts as semiconfining layer and impedes the vertical movement of water from the upper to the lower aquifer and vice versa.

The shallow and deep piezometers represent the behaviour of the water in the shallow and in the deep aquifers respectively and indicate two kinds of static water levels in the same locality. The piezometric static water level of the deep aquifer is higher than that of the shallow aquifer in the majority of the wells due to the artesian phenomena of the deep aquifer. Both levels show seasonal fluctuations. Observations indicate that the annual variation for the shallow aquifer is about 0.3 m, whereas the deep aquifer varies between 12 and 36 m from spring to summer. Fluctuations in the deep aquifer are caused by discharge from pumping while fluctuations in the shallow aquifer are caused by evaporation by capillary on the top of the aquifer, evapotranspiration in Sebkas and salty lakes during summer

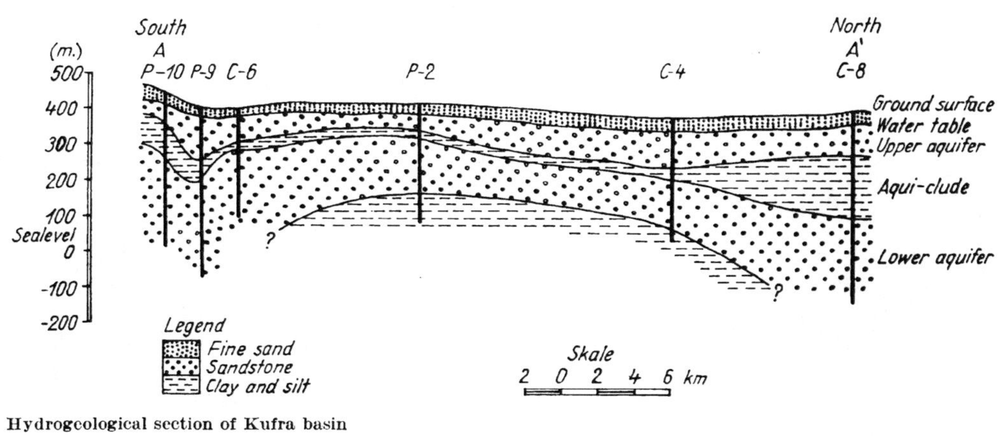

Fig. 2. Hydrogeological section of Kufra basin

time and by varying, but minor, recharge of water from the irrigation of the farm areas [2].

The hydraulic conductivity of the aquifer in the region (shallow and deep) ranges from 0·147 to 8·42 m/d; storage coefficients vary between 3×10^{-3} to 8×10^{-3} whilst transmissibility is estimated to be between 14 and 18 m^2/h [3]. Tracer experiments [3] as well as water level records [2] showed that there is no hard evidence of connection or leakage from the shallow aquifer into the deep aquifer. Moreover, chemical analysis of groundwater indicated two different water types [6]. The shallow aquifer is sodium chloride type water while the deep aquifer is sodium bicarbonate type water. Water salinity of the shallow aquifer is twice that of the deep aquifer.

Experimental Work and Results

Thirtynine water samples were collected from different localities in Kufra region representing both the shallow and the deep aquifers and one from the surface salty lake (Fig. 1). The samples were collected after sufficient water was pumped out from the recognized source to ensure complete mixing and homogeneity.

Deuterium and oxygen-18 were measured using a "Micromass 602-C" mass spectrometer with an accuracy of $\pm 1‰$ and $\pm 0{\cdot}03‰$ for deuterium and oxygen-18 respectively. Their results are expressed as $\delta‰$ values relative to SMOW. Tritium was measured in the water samples using a low-level Packard liquid scintillation spectrometer after a prior enrichment step in a special electrolysis device. Tritium measurements are tabulated as tritium units (T. U.) and the experimental error limit is given as 1σ. The radiocarbon was measured relative to the total carbon content of the sample by means of gas proportional counter using purified carbon dioxide gas as the filling gas. The average value for the ^{14}C content of recent prethermonuclear

Tab. 1. Tritium, deuterium, oxygen-18 and radiocarbon contents in Kufra groundwater

Serial	Sample		Tritium T. U.	Deuterium (δ, ‰)	Oxygen-18 (δ, ‰)	^{14}C content % modern	^{14}C age years
	A-Deep aquifer						
1	W-3,	production well	5 ± 2	−85	−11·2	0·80 ± 0·3	37,500 ± 2300
2	C-11	production well	4 ± 2	−86	−10·5		
3	C-15,	prodcution well	3 ± 2	−84	−10·6		
4	C-22,	production well	4 ± 2	−85	−11·2		
5	C-36,	production well	3 ± 2	−85	−11·0	1·90 ± 0·26	31,200 ± 1000
6	C-40,	production well	5 ± 2	−87	−11·2		
7	C-43,	production well	4 ± 3	−91	−10·8		
8	V-47,	production well	2 ± 2	−83	−11·3		
9	C-53,	production well	3 ± 3	−86	−10·9		
10	C-67,	production well	5 ± 3	−86	−11·0		
11	C-75,	production well	4 ± 3	−87	−11·4		
12	C-78,	production well	4 ± 3	−84	−11·1		
13	C-81,	production well	3 ± 3	−85	−11·2		
14	C-82,	production well	2 ± 2	−89	−10·7	0·84 ± 0·23	37,100 ± 2500
15	C-85,	production well	3 ± 2	−92	−10·9		
16	C-95,	production well	4 ± 3	−95	−11·2		
17	C-115,	production well	4 ± 3	−90	−11·3		
18	C-119,	production well	4 ± 3	−90	−11·7		
19	C-127,	production well	5 ± 3	−91	−11·1		
20	C-133,	production well	4 ± 3	−87	−11·0		
21	HQ-1,	production well	5 ± 3	−81	−11·1	0·81 ± 0·22	37,200 ± 2200
22	P-2244,	Piezometer	4 ± 3	−86	−11·1		
23	P-2211,	Piezometer	7 ± 5	−93	−11·0		
24	P-2216,	Piezometer	8 ± 5	−84	−10·6		
25	P-27,	Piézometer	5 ± 3	−84	−9.7		
26	P-28,	Piezometer	6 ± 4	−84	−10·5		
27	P-31,	Piezometer	6 ± 4	−83	−10·6		
28	P-32,	Piezometer	8 ± 6	−79	−8·9	1·82 ± 0·25	32,400 ± 1100
29	P-33,	Piezometer	8 ± 7	−79	−9·1		
30	P-34,	Piezometer	4 ± 3	−99	−10·5		
	B-Shallow aquifer						
31	P-2211,	Piezometer	11 ± 7	−84	−11·7	3·50 ± 1·10	25,600 ± 2800
32	P-2216,	Piezometer	13 ± 7	−82	−10·2		
33	P-27,	Piezometer	10 ± 8	−75	−9·0		
34	P-28,	Piezometer	12 ± 8	−78	−10·5		
35	P-31,	Piezometer	9 ± 7	−78	−9·8	2·30 ± 0·30	28,800 ± 1000
36	P-32,	Piezometer	12 ± 8	−79	−10·7		
37	P-33,	Piezometer	13 ± 9	−77	−9·3		
38	P-34,	Piezometer	10 ± 7	−92	−9·9	2·70 ± 0·20	27,700 ± 600
	C-Salty lakes						
39	East lake		22 ± 9	−21	+5·0		

waters was taken as 85%. ^{14}C ages were corrected using the different $^{13}C/^{12}C$ ratios of biogenic carbon dioxide and limestone. The ^{13}C content was measured by mass spectrometer. The results are given in Tab. 1.

Discussions and Conclusions

The interpretations of variations in the isotopic composition of groundwater are of particular importance in studies of mechanism, rate and periods of groundwater recharge. The varying environmental stable isotopic composition of natural waters is due to the fractionation processes occuring in the hydrological cycle while that of the environmental radioisotopes as tritium and radiocarbon is due to cosmic radiation and detonation of thermonuclear devices.

Tritium measurements in groundwater is a useful tool to show very modern recharges result from atmospheric contamination. Presence of tritium in groundwater indicates that a considerable amount of recharge took place within the last thirty years (age of thermonuclear tests). Absence of tritium means older contemporaneous recharge.

The deep aquifer at Kufra shows low tritium content with an average value of 4 T. U. which is near to the sensitivity of the counting system. The shallow aquifer indicates some what higher tritium values with an average of about 10 T. U. The interpretation of these figures on the bases of recharge from water of meteoric origin is difficult because of two reasons. The first is that these figures are near to the limit of sensitivity of the counting system. The second is that there is no any possible recent recharge in the area as rain is very scarce and no surface water flows around. The only explanation of the increase of tritium in the shallow aquifer is due to either differences in sampling technique or infiltration of some irrigation water, exposed to atmospheric moisture containing higher tritium values, to the top of the shallow aquifer. This is frequently observed when intensive spray irrigation is used in sandy soils and water table level is near to the ground.

The deuterium and oxygen-18 contents of groundwater could be used to differentiate between waters subjected to different physical processes. The condensation of water vapour in the atmosphere is considered to be an equilibrium process. Thus the deuterium and oxygen-18 content of the precipitated water is dually related to the residual vapour fraction. The isotopic composition of the precipitated water is a function of the temperature at the time of condensation and this temperature could be calculated according to *Dansgaard* [7] as follows:

$$\delta^{18}O = 0.695\, t_a - 13.6\,‰ ;$$

and

$$\delta D = 5.6\, t_a - 100\,‰ .$$

A parameter of interest is defined as the "*d*-parameter" [8] and expressed by:

$$d = \delta D - 8\delta^{18}O\ ‰$$

It is used to differentiate between meteoric and paleowaters ($d < 10‰$ characterized meteoric water). It was also found by *Craig* [9] that the variation of the deuterium and oxygen-18 contents in precipitation are linearly correlated by the relationship:

$$\delta D‰ = 8\delta^{18}O‰ + 10$$

The isotopic composition of water in the aquifer does not change and is thus related to that of precipitation in recharge region at the time of recharge.

The results of deuterium and oxygen-18 measurements given in Tab. 1 showed simillarity between the isotopic composition of the deep production wells and that of the deep piezometers. This indicates close relationships between water of the two sets of wells. The isotopic composition of the shallow aquifer differs from that of the deep aquifer. Oxygen-18 and deuterium contents of the deep aquifer has an average values of $-10.8 \pm 0.6‰$ and $-86 \pm 4‰$ respectively. The corresponding values for the shallow aquifer are $-10.0 \pm 0.6‰$ and $-81 \pm 5‰$. The data from these two groups are statistically distinct. The deep aquifer is more deplated in its stable isotopic composition than the shallow aquifer which confirms the disconnection between the two aquifers.

Inserting these data into Dansgaard equations, the mean annual air temperature at the time of formation of these waters was to be in the range of +2.5 °C and +5 °C for the deep and shallow aquifers respectively. This low temperature of formation is about similar to that of the present day central European mean annual air temperature but very much lower than that of the present day Saharian climate. This indicates that Kufra groundwaters were recharged during much cooler climatic conditions.

The stable isotope data given in Tab. 1 are plotted on the conventional δ diagram in which the δD values are plotted versus $\delta^{18}O$ values, Fig. 3. The meteoric line which represent the isotopic composition of present day precipitation is represented as line I. The diagram is defined by least squares fit of the recognized data.

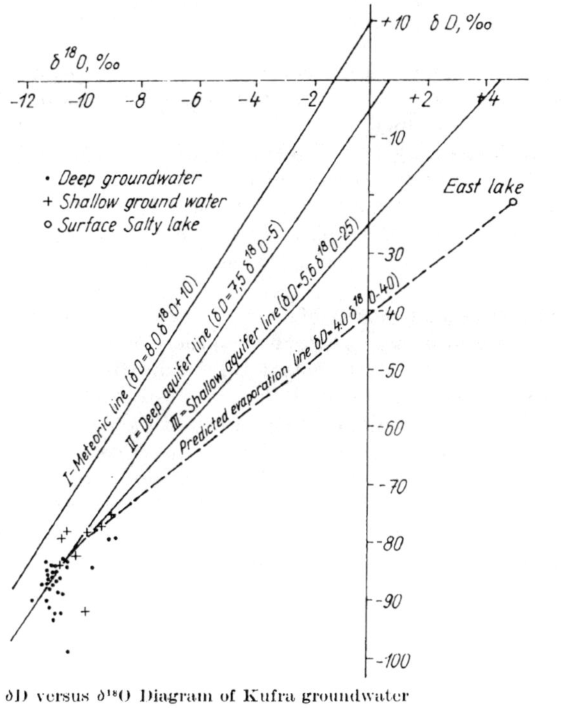

Fig. 3. δD versus $\delta^{18}O$ Diagram of Kufra groundwater

Kufra deep aquifer water is represented by line II. The best fit of this line is plotted according to the equation:

$$\delta D = 7.5\delta^{18}O - 5$$

The shallow aquifer water is represented by line III having the equation:

$$\delta D = 5.6\, \delta^{18}O - 25$$

The deep aquifer points fall on straight line parallel to the meteoric line. The "*d*-parameter" is calculated to be

$\pm 5^0/_{00}$ with small variance from the regression line which may be due to slight kinetic evaporation effect. The "*d*-parameter" is usually assumed to be due to kinetic isotope fractionation and decreases with decreasing moisture deficit of the air [10]. This means that the moisture deficit at the time of formation of Kufra groundwater was smaller than the present day value. Comparing the obtained value of the "*d*-parameter" of Kufra groundwater with the corresponding value of the present meteoric water cycle "+10‰", confirms the paleo-temperature origin of Kufra groundwater.

The decrease in the slope of the shallow aquifer indicates possible evaporation processes and both waters differ from the present meteoric water. The difference in stable isotopic composition is most probably due to an age effect. Depletion in the stable isotopic composition suggested that these groundwater mainly have been formed by local rainfall during humid phases of the past in a different meteorological patterns than that of the present day operated during much more cooler climatic conditions.

It is clear, from Fig. 2, that the shallow aquifer possesses higher isotopic composition than that of the deep aquifer. While the deep aquifer did not suffer from apparant isotopic fractionation or mixing processes at the time of condensation, the shallow aquifer suffered from some isotopic fractionation by evaporation. The predicted isotopic evaporation line in the area was constructed using the isotopic composition of a Saharian open lake in the area "East lake" and those of the groundwater. This line is found to have the equation

$$\delta D = 4.0\,\delta^{18}O - 40$$

It shows evaporation kinetics on the stable isotopic composition of groundwater in the area under the prevailing climatic conditions.

An interesting case is the piezometer, P-34, having ^{18}O content within the range of other waters while its deuterium content is much smaller than the average value. This trend is noticed in both shallow and deep measurements. This phenomena should be investigated thoroughly in the future.

Radiocarbon dating of some samples representing the shallow and deep aquifers in the area yielded ages ranging from 25,600 to 37,500 years B.P. indicating old groundwater recharge probably during the late Pleistocene. The shallow aquifer shows relative modern recharge in comparison with the deep aquifer. This is again confirms disconnection of both aquifers. Such differences in age are partly depending on differing recharge mechanisms and periods at the recharge areas and periods and partly they are the result of mixture between younger and older waters.

To explain groundwater formation in Kufra region on the bases of radiocarbon dating and environmental isotope analyses [10, 11], we have to consider that this water is fossil water without noteworthy recharge under present climatical conditions. During the formation time span, the area received sufficient winter rain from the western drift air masses carrying water from the Atlantic ocean, producing groundwater by infiltration through soil layers. This is called the great pluvial, during which most of the deep groundwaters in the central Sahara region are formed and covered the period from 30,000 to 45,000 Y.B.P. (years before present). Following this, subpluvial periods of less rains and more warm climate covered the period from 20,000 to 30,000 Y.B.P. and responsible for some younger and shallower groundwater in the southern part of the Sahara. This was followed by a period of cold and dry semi-arid climate in the age interval between 20,000 to 14,000 Y.B.P. Groundwater recharge under these semi-arid conditions was very low.

During the time span between 14,000 to 10,000 Y.B.P., heavy climate variations with winter rains from the western drift existed. This climate is then changed to a warm post glacial period and characterized by rains carried by winds from convective regions. This is the post-pluvial ages which appeared as scattered phases and decreases in the Sahara from south to east.

Younger groundwaters had probably been formed during the intermittend post-pluvial humid phases with ages ranging from some thousands to about 12,000 Y.B.P.

Received April 6, 1983

References

[1] *E. M. El-Shazly, M. Abdel-Hady, A. M. Shata*, Proc. U.N. Conference on Desertification, A/CONF. 74/24, Nairob, Kenya, 1977

[2] *A. A. Eddib*, M. Sc. Thesis, Ohio Univ. U.S.A. 1973

[3] *A. M. Eskangi, F. M. Swailem*, "Peaceful uses of Atomic Energy for Scientific and Economic Development", Proc. Symp., Baghdad, Iraq, (1975) 517

[4] *D. W. Greenman*, Report, Timpton and Kalmbach, Inc., Libya 1972

[5] *E. P. Fisk*, Report, Occidental Oil Company of Libya, Inc. 1970

[6] *A. M. Eskangi, F. M. Swailem, M. S. El-Manharawy*, "Trace Elements in Drinking Water, Agriculture and Human Life", Proc. Symp., Cairo (1977) 234

[7] *W. F. Dansgaard*, Tellus **16** (1964) 436

[8] *J. R. Gat, A. Issar*, Geochimica and Cosmochimica Acta **38** (1974) 1117

[9] *H. Craig*, Science **133** (1961) 1702

[10] *C. Sonntag, E. Klitzsch, E. P. Löhnert, K. O. Münnich, Chr. Junghars, U. Thorweihe, K. Weistroffer, F. M. Swailem*, Proc. "IAEA — UNESCO" Symp. on Isotope Hydrology, SM-228/28, Neuherberg, W.-Germany (1978) 569.

[11] Arid Zone Hydrology: Investigations with Isotope Techniques., IAEA, STI/PUB/ (1980) 547

Isotopenpraxis 20 (1984) 2, S. 69 bis 70

Kurze Originalmitteilungen
The Study of Adsorption for ^{99m}Tc Generator

Lj. M. Jaćimović, S. M. Milenković, J. L. Vučina

(The Boris Kidrič Institute of Nuclear Sciences, Vinča, Institute for Radioisotopes)[1])

Keywords

adsorption; aluminium oxides; molybdenum; radioisotope generators; technetium 99m

Introduction

The adsorption of molybdate ions on alumina depends on pH which affects changes in its structure as well as in the surface electrical charge of alumina. In acid media the formation of polymolybdate ions occurs. This process also depends, apart from pH, on the concentration of molybdenum in solution and can be formulated [1] as:

$$n\, MoO_4^{2-} + (2\,n - 2)\, H^+ \rightleftharpoons [MoO_{2n+2}(OH)_{2n-2}]^{2-}$$

The polymerization factor n depends on pH and the concentration of molybdenum. The decrease of pH and increase of molybdenum concentration result in a higher polymerization factor, thus leading to increased adsorption. Polymerization starts at $pH < 7$ and molybdenum concentration above about 0·001 M [2, 3].

The adsorption of molybdenum on alumina and determination of its adsorption capacities under conditions of low molybdenum concentration are of particular interest in the production of ^{99m}Tc generators [4, 5] based on adsorption of ^{99}Mo on alumina. However, in the praxis it is known that with the increased adsorbed ^{99}Mo activities, the elution yield of ^{99m}Tc decreases. This effect is caused by the action of different and very complex reactions induced by radiation.

Several ways of overcoming this problem were proposed, e. g. addition of some substances to alumina, addition of hydrogen peroxide to the eluent saline solution, etc. In our previous papers we have discussed the use of the mixture of alumina and silica gel in the production of ^{99m}Tc generators [6–8].

In this paper the effect of molybdenum concentration (0·01 M), liquid feed rate and pH on the adsorption characteristics of alumina were investigated. These characteristics were also determined for the proposed adsorbent consisting of a mixture of alumina and silica gel. The effect of hydrogen peroxide on adsorption capacities was also investigated.

Experimental

Adsorption of molybdenum on alumina (Woelm, acid, 100–200 mesh) was investigated and 1 g of this adsorbent was used in the experiments. A mixture of 1 g of this alumina and 0·5 g silica gel (Merck, 100–200 mesh) was also used. Determination of the adsorption characteristics of these two substrates for molybdenum was carried out under dynamic conditions. The column technique was employed. Glass columns (8 mm in dia.) were used throughout the experiments. After filling, the columns were washed with 100 ml of water. The molybdenum solution was prepared by dissolving p. a. MoO_3 (Merck) in 4 N NaOH. The desired pH of the solution was adjusted with diluted hydrochloric acid. The solution of the fission product ^{99}Mo (IRE-Belgium or ZFK-GDR) was added as a radiotracer so that the ^{99}Mo activity was 55·5 kBq/ml. The final solution was 0·01 M in molybdenum. This solution was in two ml portions passed through the columns at flow rates of 8, 20 and 120 (ml min^{-1} cm^{-2}). The breakthrough capacity of each substrate was determined by measuring the ^{99}Mo activity in the effluent. Saturation was determined by measuring the ^{99}Mo in the effluent up to the point when the activity per ml of effluent was identical to that per ml of influent solution. Experiments with hydrogen peroxide were carried out using the same technique. Previously 3 ml of hydrogen peroxide (12%) were passed through the columns which were then washed with 50 ml of water.

^{99}Mo activity was measured by an automatic ICN Gamma Counter (Belgium).

[1]) Address: 11001 Beograd, P.O. Box 522, Yugoslavia

Results and Discussion

Results of the determination of the adsorption characteristics of alumina for molybdenum are summarized in Tab. 1.

Tab. 1. Adsorption characteristics of alumina
Adsorbent: 1 g alumina (Woelm, acid, 100 to 200 mesh)

pH of the column	^{99}Mo solution flow rate ml min^{-1} cm^{-2}	Breakthrough capacity mg Mo/g	Saturation capacity mg Mo/g
1·5	8	30·6	71
	20	30·6	72
3	8	24·9	49·5
	20	21·1	53·5
4	8	17·2	52
	20	17·2	48
5	8	9·6	41
	20	9·6	36
6	8	5·7	8·5
	20	5·7	9

Tab. 2. Comparison of adsorption characteristics of alumina and alumina in a mixture with silica gel
Adsorbents: 1 g alumina (Woelm, acid, 100–200 mesh)
1 g alumina + 0·5 g silica gel (Merck, 100–200 mesh)
^{99}Mo solution flow rate: 20 ml min^{-1} cm^{-2}
Mo concentration: 0·01 M

*p*H of the column	Breakthrough capacities (mg Mo/g)			
	Al_2O_3	$Al_2O_3 + H_2O_2$	$Al_2O_3 + SiO_2$	$Al_2O_3 + SiO_2 + H_2O_2$
1·5	30·6	9·6	30·6	7·7
3	21·1	21·1	17·2	15·3
4	17·2	15·3	17·2	15·3
5	9·6	9·6	9·6	9·6
6	5·7	5·7	3·8	3·8

*p*H of the column	Saturation Capacities (mg Mo/g)			
	Al_2O_3	$Al_2O_3 + H_2O_2$	$Al_2O_3 + SiO_2$	$Al_2O_3 + SiO_2 + H_2O_2$
1·5	72	55·5	70	55
3	53·5	52·5	54	53
4	48	50·5	50	46
5	36	40	36	40
6	9	16	16	18

From the data in Tab. 1 the dependance of molybdenum adsorption on *p*H can be seen. The breakthrough capacities from 30·6 mg Mo/g at *p*H = 1·5 fall to only about 5 mg Mo/g at *p*H = 6. The saturation capacities behave similarly. At *p*H = 6 saturation is already achieved at 9 mg Mo/g. It was found that the used flow rates at which molybdenum solution is passed through the columns do not have a decisive effect on the adsorption characteristics of alumina. Practically the same results were obtained for both flow rates used. In some experiments high flow rates of about 100 (ml min^{-1} cm^{-2}) were used, but these flow rates were found to be unsuitable.

In Tab. 2 data on the adsorption characteristics of alumina and an adsorbent consisting of a mixture of alumina and silica gel are compared. The effect of hydrogen peroxide on these characteristics is also included. A flow rate of 20 (ml min^{-1} cm^{-2}) was chosen for the experiments.

From the data in Tab. 2 it can be concluded that the adsorption characteristics of alumina itself and of its mixture with silica gel are practically the same. The breakthrough and saturation capacities in the investigated *p*H region do not differ considerably under the given set of experimental conditions.

Treatment of the adsorbents with hydrogen peroxide before adsorption seems to have a significant effect on the breakthrough capacities but only in higher acid media. The breakthrough capacities are about 3 times lower when compared with the untreated columns. The saturation capacities at *p*H = 1·5 are also decreased but this effect is not so pronounced. The saturation capacities at *p*H = 6 are higher in comparison with the untreated columns.

Conclusion

Examination of the adsorption characteristics of alumina (Woelm, acid) for molybdate ions in acid media shows that the capacities are dependant on *p*H. The breakthrough and saturation capacities are about 30 mg Mo/g and 70 mg Mo/g, respectively, at *p*H = 1·5 but only 4–6 mg Mo/g at *p*H = 6. The addition of silica gel does not alter the adsorption characteristics of the obtained mixture. Treatment of the adsorbents with hydrogen peroxide before adsorption greatly decreases the breakthrough capacities at *p*H = 1·5. The saturation capacities are also decreased but to a lesser extent. It was found that the flow rates of molybdenum solution in the range of 10–20 (ml min^{-1} cm^{-2}) do not affect the adsorption characteristics of the investigated sorbents. These results confirm that the proposed adsorbent consisting of the mixture of alumina and silica gel can be successively used in the production of ^{99m}Tc generators.

Received June 15, 1983

References

[1] *K. Yr. Shapiro, V. V. Kulakova, E. D. Evstigneeva, V. N. Zuev, L. A. Nenasheva*, Zh. Neorg. Khim. **15** (1970) 2238

[2] *E. Pungor, A. Halasz*, J. Inorg. Nucl. Chem. **32** (1970) 1187

[3] *Yu. V. Morachevskii, L. I. Lebedeva*, Zh. Neorg. Khim. **10** (1960) 1165

[4] *H. Arino, H. H. Kramer*, Int. J. Appl. Radiat. Isotopes **26** (1975) 301

[5] *S. M. Milenković, J. L. Vučina, Lj. M. Jaćimović, E. S. Karanfilov, T. V. Memedović*, Isotopenpraxis **19** (1983) 85

[6] *J. L. Vučina, Lj. M. Jaćimović, S. M. Milenković*, IV Int. Symp. on Radiopharmaceutical Chemistry, Jülich (FRG) 1982, p. 298

[7] *S. M. Milenković, J. L. Vučina, Lj. M. Jaćimović, E. S. Karanfilov, T. V. Memedović*, VIIth Int. Symp. on Nuclear Medicine, Karlovy Vary (Czechoslovakia) 1983, p. 54

[8] *S. M. Milenković, J. L. Vučina, Lj. M. Jaćimović, E. S. Karanfilov, T. V. Memedović*, VIIth Int. Symp. on Nuclear Medicine, Karlovy Vary (Czechoslovakia) 1983, p. 55

Isotopenpraxis 20 (1984) 2, S. 71 bis 73

Получение биомассы *хлореллы*, обогащенной ^{14}C-крахмалом

В. Л. Карпов, А. Н. Деденков

(Научно-исследовательский институт медицинской радиологии АМН СССР, г. Обнинск)[1])

Keywords

algae; biosynthesis; carbon 14; labelled compounds; starch

Общемеченые ^{14}C-крахмал и глюкозу обычно получают методом биосинтеза с использованием листьев высших растений [1, 2]. Недавно появилось сообщение *Доуха* и соавторов [6] с использованием для подобных целей одноклеточных водорослей — организма более удобного, чем высшие растения. В этой работе специальный штамм Chlorella vulgaris культивируется на $^{14}CO_2$ в присутствии ингибиторов белка. В результате получается ^{14}C-крахмал с высокой удельной активностью и выходом по отношению к общей радиоактивности клеток равным 50%.

В данной работе описывается метод получения биомассы хлореллы, содержащей 70% ^{14}C от общей радиоактивности клеток и удельной активностью равной 80% от удельной активности исходного сырья. Изменение биосинтеза в сторону образования крахмала достигнуто путем культивирования водорослей на среде без минерального азота.

В химическом составе биомассы хлореллы, выращенной на полной питательной среде, преобладают белки, составляющие 40—60% от сухого веса. При исключении из среды соединений азота биосинтез у одних штаммов хлореллы изменяется в сторону образования липидов, у других — углеводов [3]. Например, у Chlorella SpK содержание крахмала к концу 7-х суток выращивания на среде без азота повышается до 55%. Во вновь образованных соединениях хлореллы доля крахмала составляет 70—80%. Исходя из этого можно ожидать, что при замене обычной углекислоты на радиоактивную при культивировании в условиях азотного голодания большая доля ^{14}C попадает в крахмал. Удельная активность будет зависеть от соотношения образовавшегося из $^{14}CO_2$ крахмала и исходного крахмала. Например, если вес приросшего на радиоактивной углекислоте крахмала вчетверо больше исходного, то удельная активность его составит 80% от удельной активности ихсодного сырья ($^{14}CO_2$). Пределы изменения соотношения прироста и исходного уровня крахмала ограничены. 55% крахмала в хлорелле достигается к концу 7-х суток культивирования на среде без азота [3]. На $^{14}CO_2$ 7-дневное культивирование невозможно из-за переоблучения культуры от β-часиц ^{14}C [4], поэтому увеличение соотношения можно достичь лишь путем снижения исходного уровня крахмала в клетках. По нашим данным при выдерживании хлореллы в темноте при 37 °C с продувкой суспензии воздухом через 2, 4, 6, 8 и 20 часов содержание крахмала в клетках снижается с 11% до 8,7, 6,4, 4,5, 7,4, 5,0% соответственно. Как видно из данных, скорость падения содержания крахмала замедляется и, по-видимому, полностью избавиться от него невозможно, а поэтому неизбежно разбавление радиоактивного крахмала. Если удовлетвориться 20% разбавлением радиоактивного углерода (прирост крахмала в клетках с 5 до 25%), то это достижимо за 20—30 часов культивирования на радиоактивной углекислоте и с выходом по радиоактивности крахмала равным 70%.

Пример переработки 100 мкюри $^{14}C_2$, образующегося из карбоната бария, в крахмал указанной удельной активности, приводится ниже. В тех случаях, когда требуется получить биомассу хлореллы, обогащенную крахмалом с предельной удельной активностью, культивировать хлореллу на радиоактивном CO_2 следует до прекращения его усвоения. В этом случае выход в расчете на исходную радиоактивность карбонатов будет ниже.

Включение ^{14}C в крахмал культурой Chlorella SpK в условиях недостатка в среде азота.

В целях отработки препаративного метода получения хлореллы, обогащенной ^{14}C-крахмалом с высокой удельной активностью, было поставлено ряд опытов, сведенных в табл. 1. Необходимо было определить с какого момента после перенесения клеток на среду без азота устанавливается максимальный биосинтез крахмала, а также влияние на этот биоситез темнового культивирования хлореллы, необходимого в свою очередь для истощения исходного уровня крахмала и тем самым ограничения разбавления ^{14}C-крахмала.

Как видно из табл. 1 при культивировании хлореллы в нормальных физиологических условиях при наличии в среде всех минеральных элементов за 4 часа фотосинтеза на радиоактивной углекислоте образуется 11% меченого крахмала. Основная радиоактивность в хлорелле, выращенной в этих условиях, находится в белках (51%) и низкомолекулярных веществах спиртового экстракта (38%).

При перенесении водорослей на среду без азота с последующим культивированием на радиоактивной углекислоте за 4 часа содержание ^{14}C-крахмала поднимается до 38% (опыт 2). Однако в этих условиях клеткам, по-видимому, еще доступен источник азота для белкового синтеза за счет внутриклеточного пула и истощать его можно путем фотосинтеза на обычной углекислоте.

[1]) Адрес: Koroljewa Str., Haus 4; 249020 Obninsk, UdSSR

Табл. 1. Содержание радиоактивного крахмала в клетках хлореллы после 4-х часового фотосинтеза на радиоактивной углекислоте

№ опыта	Условие проведения фотозинтеза	радиоактивность в %		
		крахмал	спиртовой экстракт	остаток
1.	Вдвое разбавленная среда Тамия	11	38	51
2.	Среда Тамия без KNO_3	38	19	43
3.	Среда Тамия без KNO_3 с предварительным 3-х часовым фотосинтезом на $^{14}CO_2$	58	26	16
4.	Среда Тамия без KNO_3 с предварительным 6-ти часовым фотосинтезом	71	21	8
5.	То же, что в опыте 4, но предварительно 20 часов культивирование в темноте	69	20	11

В опыте 3 хлорелла, перенесеннная на среду без азота перед ,,горячим‘‘ фотосинтезом, 3 часа выдерживалась на свету на обычной углекислоте. Включение радиоуглерода в крахмал в этих условиях повысилось до 58%, а после 6-ти часового предварительного культивирования включение радиоуглерода в крахмал поднялось до 71%. Практически не влияет на направленность биосинтеза выдерживание хлореллы в темноте при 37 °C с продуванием воздуха в течении 20 часов (опыт 5). Таким образом, максимальный биосинтез крахмала начинается спустя 3-6 часов после перенесения культуры на среду без азота. При проведении препаративного биосинтеза в течение 24 часов значение пониженного синтеза в первые часы после перенесения культуры на среду без азота будет незначительно и преинкубирование на свету на обычной углекислоте может быть опущено или сокращено до 1 часа. Истощение внутриклеточного пула соединений минерального азота, по-видимому, возможно и в темноте, когда крахмал не образуется, но мы этого не проверяли.

Как уже отмечали, культивирование хлореллы в темноте приводит к уменьшению исходного уровня крахмала, но мы не уверены, возможно ли полностью от него избавиться и таким путем свести к минимуму разбавление тяжелого изотопа в крахмале легким. В случае использования данного метода получения крахмала меченого стабильным (^{13}C) изотопом, а через него и многих других соединений, например, равномерно меченых аминокислот с помощью соответствующих гетеротрофных продуцентов, разбавление легким изотопом углерода может быть сведено до 10%.

Экспериментальная часть

Методика

В работе использовали зеленые одноклеточные водоросли штамм Chlorella SpK. Фотосинтез в атмосфере $^{14}CO_2$ проводили при 37 °C в герметичной установке, описанной ранее [5]. Циркуляцию радиоактивной углекислоты по системе и перемешивание культуры хлореллы проводили с помощью микрокомпрессора. В качестве осветителей служили люминесцентные лампы, дающие около 14 тыс. люкс (48 тыс. эрг/см^2 в сек) с каждой стороны. За счет осветителей поддерживалась температура водорослей. Избыток тепла снимался автоматическим включением воздушного вентилятора. Культиватор изготовляли из оргстекла с плоскопараллельными стенками.

Азотное голодание создавали путем перенесения осадка отцентрифугированных клеток в раствор, состоящий из вдвое разбавленной среды Тамия без KNO_3. Эту операцию повторяли дважды. Исходная концентрация клеток для фотосинтеза на радиоактивной углекислоте составляла 140 млн. клеток/мл.

Содержание ^{14}C-крахмала в биомассе хлореллы определяли после гидролиза осадка водорослей предварительно обработанных 70% метанолом, метанолом и эфиром, 2 N соляной кислотой при 100 °C в течение 2-х часов. Гидролизат разбавляли в 10 раз водой и пропускали через колонку катионита Дауэкс 50 в H-форме объемом 5 мл. Радиоактивность фильтрата определяли на жидкостном сцинтилляционном счетчике. Более 90% радиоактивности фильтрата, по данным бумажной хроматографии, содержится в глюкозе. Измеряли также радиоактивность спиртового экстракта и остатка после гидролиза клеток. В первом содержатся низкомолекулярные вещества клеток: углеводы, свободные аминокислоты, хлорофил, липиды и другие вещества. В остатке после гидролиза содержатся белки, клеточные стенки и другие полимерные соединения.

Пример препаративного биосинтеза

Расчет исходного количества клеток. При эпизодических синтезах меченого крахмала обычно требуется, чтобы радиоактивная углекислота была использована водорослями полностью и при этом удельная активность конечного продукта получилась максимальной. Как уже отмечалось выше, удельная активность ^{14}C-крахмала зависит от соотношения между приросшим на радиоактивной углекислоте крахмалом и крахмалом находящимся в инокулюме. От исходного количества крахмала зависит исходное количество клеток хлореллы. В описываемом ниже примере стояла задача получить биомассу хлореллы, обогащенную крахмалом с удельной активностью последнего около 80% от удельной активности исходного $^{14}CO_2$. Предстояло переработать 100 мкюри $Ba^{14}CO_3$ весом 500 мг. Путем расчета находим, что из углерода карбоната бария может быть синтезировано 72 мг крахмала. Учитывая, что в крахмал попадает не более 80% углерода карбоната бария, привес крахмала составит 57,6 мг. Принимая, что при культивировании в течение суток в темноте уровень немеченого крахмала в хлорелле составит 5% от сухого веса клеток и эти 5% должны весить 20% от ожидаемого привеса крахмала т. е. 11,52 мг, исходный сухой вес клеток должен быть 11,52:0,05 = 230 мг.

Вес 10^9 клеток используемого штамма хлореллы равен 6,7 мг, следовательно в 230 мг будет 34,3 . 10^9 клеток.

Порядок работы. Культуру клеток хлореллы, находившуюся в экспоненциальной фазе роста, поместили в темноту при 37 °C и в течение суток продували через нее стерильный воздух. Затем клетки отделили от жидкости центрифугированием при 2000 g, развели половинной средой Тамия без азота, снова центрифугировали. Эту операцию повторили дважды, и развели той же средой до концентрации 140 млн клеток в мл. 245 мл этой суспензии поместили в 500 миллилитровую камеру с плоскопараллельными стенками, имеющими просвет 20 мм, и в течении часа культивировали на 2% обычной углекислоте. После этого камеру с водорослями и всю систему продули обычным воздухом и подключили резервуар с 5% радиоактивной углекислотой. В процессе культивирования регистрировали уровень радиоактивности в газовой фазе. К 23 часам уровень радиоактивности снизился почти до фона, суспензию подкислили до pH 4, подключили поглотитель CO_2 и поглотили оставшуюся радиоактивную углекислоту. Клетки отделили от жидкости, извлекли хлорофил и другие низкомолекулярные вещества метанолом и бесцветный остаток клеток кипятили в течении 2-х часов в 2 N HCl для гидролиза крахмала. Выход глюкозы-сырца составил 69 мкюри, удельная активность —187 кюри/моль.

Поступила в редакцию 8/VI, 1983 г.

Литература

[1] *С. Аронов.* Изотопные методы в биохимии. Из-во ИЛ. М. 1959, стр 204

[2] *О. Д. Быков.* Труды Петергофского Биологического Института ЛГУ II 18 (1960) 68

[3] *Т. С. Жукова, Г. Л. Клячко-Гурвич, М. Г. Владимирова, Т. А. Курносова.* Физиология растений **15** (1968) 96

[4] *В. Л. Карпов, Б. В. Филиппов.* Радиобиология **5** (1965) 580

[5] *В. Л. Карпов, О. Д. Быков, Ю. М. Селезнев.* Физиология растений **16** (1969) 169

[6] *И. Доуха, И. Колина, И. Щетлик, М. Смажик,* 2-й симпозиум стран-членов СЭВ. Огранические соединения меченые радиоактивными изотопами (1981). Тезисы докладов, стр 50

Isotopenpraxis 20 (1984) 2, S. 73 bis 74

Zur Stabilität von [methyl-³H] Thymidin-5′-triphosphat

T. Brankoff

(Akademie der Wissenschaften der DDR, Zentralinstitut für Kernforschung Rossendorf, Bereich III)[1])

Keywords

autoradiolysis; labelled compounds; thin-layer chromatography; thymidine nucleotides; tritium compounds

Die primären Zersetzungsprodukte eines Nucleosidtriphosphats sind unter neutralen Bedingungen das entsprechende Nucleosiddiphosphat und anorganisches Phosphat. Später treten dann Nucleosidmonophosphat, Nucleosid bzw. Base auf. Um die autoradiolytische Zersetzung eines tritiummarkierten Nucleosidtriphosphats, bedingt durch die β-Strahlung, so gering wie möglich zu halten, wird die Lagerung in 50%igem Ethanol unter Zusatz von 1% Ammoniumbikarbonat, mit maximal 37 MBq/ml radioaktiver Konzentration und unter Kühlung empfohlen [1]. Von tritiummarkiertem Thymidin-5′-triphosphat (TTP) wurde die Stabilität bei unterschiedlichen Lagertemperaturen, vor allem bei −140 °C und −196 °C bestimmt [1−4]. Wir untersuchten die Stabilität bei −25 und −5 °C.

Für die Untersuchungen wurde [methyl-³H]TTP [5] verwendet, das wir aus [methyl-³H]Thymidin [6] durch chemische Phosphorylierung herstellten. Die Reinigung von [³H]TTP erfolgte über eine DEAE-Sephadex-Säule mit Triethylammoniumbikarbonat und anschließende Zersetzung dieses Puffers in Gegenwart von Ethanol im Vakuum [5]. Dieses [³H]TTP wurde in 50%igem Ethanol ohne weiteren Zusatz und ohne Sterilisation bei −25 °C oder −5 °C gelagert.

Als Analysenmethode wurde die Radio-Dünnschichtchromatographie verwendet. Die Trennung von Nucleosidphosphaten auf Zellulose- sowie Ionenaustauscher-Zellulose-Schichten ist aus der Literatur [7, 8] bekannt. Zur Trennung des [³H]TTP von seinen Radiolyseprodukten wurde von uns als Adsorbens PEI-Zellulose F und als Fließmittel Ammoniumbikarbonat analog [2, 9] benutzt.

Für die Stabilitätsuntersuchungen wurden 6 [³H]TTP-Präparate von ROTOP mit spezifischen Aktivitäten von 500−650 GBq/mMol verwendet. Die Präparate wurden mit der radiochemischen Konzentration von rund 40 MBq/ml 50%iges Ethanol (spezial rein, Apolda) bis zu einem Jahr gelagert, davon 4 Proben bei −25 °C und 2 andere bei −5 °C. In Abständen von 2−3 Monaten wurden jeweils 3 Dünnschichtchromatogramme auf PEI-Zellulose-Platten (Merck) mittels 0,2 M Ammoniumbikarbonat angefertigt. Für die tritiummarkierten Substanzen wurden die R_F-Werte erhalten: TTP = 0,20, Thymidin-5′-diphosphat = 0,45, Thymidin-5′-monophosphat = 0,70, Thymidin = 0,85.

Die radiochemische Reinheit der Proben wurde durch anschließendes Radioscanning am Berthold-Dünnschichtscanner LB 2723 II, mit angeschlossenem Drucker, ermittelt (Aktivitätsbereich 0,2−0,4 MBq, Blendenbreite: 1 mm, Scangeschwindigkeit: 300 mm/h, Integrationszeit: 1 Wert je 10 s).

Die Auswertung der Meßreihen erfolgte mittels linearer Regression. Die radiochemische Reinheit der Präparate kann durch die Beziehung $c(t) = c_0 - at$ dargestellt werden. Dabei ist c_0 − die radiochemische Reinheit am Anfang, $c(t)$ − die radiochemische Reinheit zur Zeit t und a − die monatliche Abnahme der radiochemischen Reinheit.

Die eingesetzten [³H]TTP-Präparate hatten kurz nach ihrer Herstellung eine radiochemische Reinheit von c_0

[1]) Anschrift: 8051 Dresden, Postfach 19

zwischen 95 und 98%. Der Korrelationskoeffizient für die erhaltenen Regressionsgeraden liegt bei 0,77 und die Standardabweichung der Meßwerte von diesen beträgt 1,5%.

Insgesamt wurde für die bei −25 °C gelagerten Präparate eine mittlere monatliche Abnahme der radiochemischen Reinheit von $a = 3{,}7 \cdot 10^{-3}$ Monat^{-1} (Standardabweichung $0{,}65 \cdot 10^{-3}$ Monat^{-1}) bestimmt. Nach einem Jahr bedeutet das durchschnittlich 4,5% Abnahme. Das ist relativ wenig. Für die bei −5 °C gelagerten Proben von anderen Chargen ergab sich für $a = 9{,}1 \cdot 10^{-3}$ Monat^{-1}, d. h. 10% Abnahme nach einem Jahr. Auf Grund dieser Ergebnisse bevorzugen wir für die Lagerung von [^{3}H]TTP −25 °C im Vergleich zu −5 °C.

Eingegangen am 2. 6. 1983

Literatur

[1] *E. A. Evans*, Tritium and its Compounds, London: Butterworth, 1974, S. 674, 707, 727

[2] *G. Sheppard, H. C. Sheppard, J. F. Stivala*, J. lab. Comp. **10** (1974) 557

[3] *E. A. Evans*, Self-decomposition of radiochemicals, Review 16, The RCC Amersham 1976, S. 56

[4] *E. A. Evans, M. Muramatsu*, Radiotracer Techniques and Applications, New York and Basel: Marcel Decker, 1977, Vol. 1

[5] *T. Brankoff, E. Mittag*, Jahresbericht ZfK-367 (1978) S. 89

[6] *U. Pleiß*, Jahresbericht ZfK-367 (1978) S. 92

[7] *E. Stahl*, Dünnschichtchromatographie. Berlin, Heidelberg, New York: Springer-Verlag 1967, S. 755

[8] *G.-J. Krauß, G. Krauß*, Experimente zur Chromatographie, Berlin: Deutscher Verlag der Wissenschaften 1979, S. 122

[9] *R. Monks, K. G. Oldham, K. C. Tovey*, Labelled Nucleotides in Biochemistry, Review 12, The RCC Amersham 1971, S. 28

Isotopenpraxis 20 (1984) 2, S. 74 bis 75

Tagungsberichte

3. Internationale Tagung „Nukleare Analysenverfahren" 11.–15. 4. 1983, Dresden

Die Tagung wurde von einem wissenschaftlichen Komitee unter Leitung von Prof. Dr. sc. *S. Niese* durch das Zentralinstitut für Kernforschung Rossendorf der AdW der DDR in Zusammenarbeit mit der chemischen Gesellschaft der DDR vorbereitet und durchgeführt. 275 Wissenschaftler, darunter 72 Ausländer aus 12 Ländern hatten Gelegenheit, sich an Hand von 8 Plenarvorträgen umfassende Darstellungen wesentlicher Teilgebiete geben zu lassen und in 165 Diskussionsvorträgen, darunter 75 Postern, neue Ergebnisse auszutauschen.

Nuklearanalytische Methoden, wie Aktivierungsanalyse, Kernstrahlungsspektrometrie, Kernreaktionsanalyse, Protoneninduzierte Röntgenstrahlung, Sekundärionenmassenspektrometrie und Indikatorverfahren werden einzeln oder in interdisziplinärer Zusammenarbeit kombiniert angewandt. Hinzu kommen nichtnukleare Verfahren, wie Atomabsorption und Röntgenfluoreszenz, die zur Lösung von Problemen der Reaktortechnik, der Brennelementherstellung oder Isotopenproduktion herangezogen werden. Dieser Mannigfaltigkeit entsprach das breite Themenspektrum, bei der Vertreter unterschiedlicher Disziplinen Fragen von prinzipieller Bedeutung wie Energie-, Rohstoff- und Materialprobleme, Umweltschutz und Gesundheitsfürsorge erörterten.

Im Mittelpunkt der Vorträge zur Kerntechnik einschließlich Aktinidenchemie stand auch diesmal die Bestimmung von Spaltprodukten und Transuranen in hochaktiven Materialien und Korrosionsproduktion. *Nebel* (ZfK Rossendorf) unterstrich in seinem Plenarvortrag die Bedeutung der radioanalytischen Betriebskontrolle in Kernkraftwerken, wobei eine in-situ Analytik des Korrosionsangriffs gefordert werden muß und ein noch höherer Automatisierungsgrad nur bei einer rascheren und direkteren Rückwirkung auf die Betriebsführung sinnvoll ist. *Lipovskij* (Leningrad, SU) zeigte in seinem Plenarvortrag die Notwendigkeit, die Genauigkeit im Abbrand ermittelten Isotopenverhältnisse wesentlich zu erhöhen und aus der Vielzahl heute angebotener Analysenmethoden die optimale zu wählen.

Eine Anzahl Untersuchungen galt der Extraktionschromatographie, insbesondere zum Purexprozeß. Berichtet wurde u. a. über kinetische Untersuchungen am Zentrifugalextraktor (*Eichler* et al., ZfK), die Erleichterung der Abtrennung von Transplutoniumelementen durch Oxydation mit $(NH_4)_2S_2O_8$ (*Füchtner*, ZfK), die Pu-Extraktion mit Di-n-octyl-sulfoxid (*Zambrana* et al., Cuba), die Trennung der Transplutoniumelemente als α-Hydroxy-butyrat-Komplex in Wasser-Dioxanmischungen (*Baraniak* u. *Nebel*, ZfK) sowie über den Einsatz der in Dubna (SU) entwickelten Hochtemperatur-Gaschromatographie mit auf Ti- und Mo-Folien aufkondensierten Ca-Filmen (*Hübener* u. *Timochin*, Dubna, SU).

U. u. *S. Niese* (ZfK) stellten vergleichende aktivierungsanalytische und α-spektrometrische ^{237}Np-Bestimmungen vor, wobei die NAA sich als um 4 Größenordnungen empfindlicher erweist.

Mit Überwachungstechniken für möglicherweise freiwerdende Radioaktivität machte *Knobus* (ZfK) bekannt, der über gammaspektroskopische Messungen bodennaher atmosphärischer Schwebestäube berichtet. *Clajus* et al. (SAAS Berlin) referierte über α-spektrometrische Messungen des ^{230}Th-Gehaltes in Umweltmedien, während *Denschlag*, (Mainz, BRD) den aktivierungsanalytischen Nachweis geringer Actinidenkonzentrationen auf Spaltspurdetektorfolien vorführten.

Die eminente Bedeutung nuklearanalytischer Methoden in den Lebenswissenschaften zeigte sich in einer großen Zahl von Vorträgen, darunter allein 4 Plenarvorträgen, zu diesem Thema. *Münze* (ZfK) gab eine Einschätzung der Verfahren, mit denen Aussagen zur chemischen Form trägerfreier radioaktiver Substanzen gemacht werden können, welche gegenwärtig im Hinblick auf Biolokalisationsmechanismen der in-vivo-Radiodiagnostika an Bedeutung gewinnen. *Schütte* (Inst. f. Biochemie d. Pflanzen, Halle) zeigte, wie einerseits durch Verabreichung geeigneter, spezifisch markierter Vorstufen der Bildungsweg der Nährstoffe im Organismus aufgeklärt werden kann, andererseits das Biotransformationsverhalten von Xenobiotica durch Zugabe des radioaktiv markierten Wirkstoffs leicht verfolgbar ist.

Herzmann und *Vormum* (ZfI, Berlin-Buch) machten in ihrem Plenarvortrag grundlegende Ausführungen über die einzelnen Subtypen der Sättigungsanalyse und stellten besonders eingehend die ^{125}I-Markierungsverfahren dar. *Stöcklin* (Jülich, BRD) stellte die Bedeutung der kurzlebigen Zyklotronnuklide heraus, die als markierte Biomoleküle die nichtinvasive, regionale Erfassung einzelner Organfunktionen in vivo gestatten, wobei er in z. T. spektakulären Beispielen externe Messungen des Herz- und Hirnstoffwechsels belegte.

Die aktivierungsanalytische Spurenbestimmung in biologischen Materialien stellt auch weiterhin ein umfangreiches Gebiet dar, wobei die Erkundung der Widerspiegelung gewisser Krankheitszustände durch vom Gesunden abweichende Spurengehalte — insbesondere mit dem Ziel epidemiologischer Untersuchungen und der Carcinomdiagnostik — im Vordergrund der Rossendorfer Arbeiten steht. (*Wiesener*, *Niese*, *Görner*). In Zusammenarbeit mit *Mende* (Med. Akad. Dresden) und *Schicke* (Klinikum Berlin-Buch) wurde der Spurenelementgehalt bei Struma untersucht, zusammen mit *Baldauf* (Klinikum Berlin-Buch) eine teilweise Signifikanz bei Malignomen gefunden, gemeinsam mit *Kauf* (FSU Jena) wurde das Normallevel während des ersten Lebensabschnitt des Menschen bestimmt. Neue Erkenntnisse zum Spurenelementgehalt bei Wundheilvorgängen trug *Lux* (München, BRD) vor.

Die weite Verbreitung nuklearer Analysenverfahren, besonders der in-situ-Verfahren, wurde eindrucksvoll von *Surkov* (Moskauk SU) illustriert, der die bisherigen Erfolge praktischer Kosmo-„Geologie" (Direktuntersuchungen an Mondmaterial sowie auf der Mond-, Venus- und Marsoberfläche) beschrieb. Das Arbeitsfeld der NAA in der Geologie reicht von Gesteinsuntersuchungen, bei denen Verteilungsmuster in Abhängigkeit von Korngröße, Chemismus und Verwitterungstyp erörtert werden (*Bobrov* et al., Novosibirsk, SU; *Hahne* et al., Zipe, Potsdam; ZfK), über die Multielementbestimmung in latheritischen Erzen (*Garcia*, Havanna, Cuba) bis zur ^{40}Ar/^{39}Ar-Altersbestimmung eines Basits der Ost-Antarktis (*Kaiser* u. *Loos*, BA Freiberg, ZfK).

Das Methodenarsenal umfaßt neben der NAA mit thermischen Neutronen sowohl den Einsatz des Neutronengenerators (*Hoang Dac Luc* et al., Hanoi, Vietnam), des Mikrotrons (*Hernandez* u. *Zamyatin*, Dubna, SU; *Gerstenberger* und *Geisler*, ZfI Leipzig), des Synchrotrons zur Röntgenfluoreszenzanalyse (*Bobrov* et al., Novosibirsk, SU) als auch die Spaltspurmethode, die als natürliche Speicherung der Spontanspaltung des ^{238}U in gewissen Gesteinen nutzbar ist (*Stahms* u. *Trettin*, ZfI Leipzig)

Die Aktivierung mit geladenen Teilchen und Photonen ist aufwendig, aber in Sonderfällen unerläßlich (*Aleksandrova*, Moskau, SU; *Lacroix* et al., Orlens, Frankreich (InP); *Böttger* et al., ZfK (Mg in Si); *Petri*, Jülich, BRD, (Ni mit p). Geht es um die Untersuchung der Oberfläche von Festkörpern, so sind Verfahren wie *Rutherford*-Rückstreuung (*Kudinov*, Moskau, SU; *Grötzschel*, ZfK), Ellipsometrie (*Lohner* et al., Budapest, Ungarn), PIXE (*Lenk* u. *Rößiger*, ZfI Leipzig), promter γ-Messung (*Rudolph* et al., *Ascheron* et al., ZfK, ZfI Leipzig) und Autoradiographieanalyse gefragt. Alle diese Verfahren spielten auf der Tagung eine beachtliche Rolle.

Bevorzugte Methode für die in-bulk und Profiluntersuchung ist die NAA (*Flachowsky*, ZfI *Leipzig*; *Ross*, ZfK).

Eine große Zahl von Vorträgen war methodischen Fragen der Aktivierungsanalyse gewidmet. Die Untersuchung von Standardmaterial stand im Mittelpunkt des Plenarvortrages von *Kosta* (Ljubljana, Jugoslawien). Interessante Ausführungen über das K_0-Monostandardverfahren wurde von *De Corte* et al. (Gent, Belgien; Budapest, Ungarn; Peking, China) gemacht. Technisch hochentwikkelte Anlagen sowohl zur Messung sehr hoher Zählraten (*Görner* u. *Richter*, ZfK) als auch sehr niedriger Zählraten (*Helbig* et al., ZfK) wurden vorgestellt. Last not least sind eine ganze Reihe von Arbeiten zu erwähnen, bei denen nuklearanalytische Methoden unmittelbar praxiswirksam werden. Stellvertretend seien genannt: Verschleißuntersuchungen an Elektromotoren (*Wagner* et al., IcT Berlin; Hüttenwerk Thale), Schnellbestimmung von Si in Kohlenstaub (*Maul*, ZfI Leipzig) oder etwa der Einsatz von Edelgas-Ionisationsdetektoren zum Nachweis von Plasmaabätzprodukten in der Mikroelektronik (*Mothes* et al., ZfI Leipzig).

Die Kurzfassungen der Vorträge sind im Beiheft 69 der „Mitteilung der Chemischen Gesellschaft der DDR" erschienen. Es ist geplant, neben der Veröffentlichung ausgewählter Vorträge in der „Isotopenpraxis", alle Beiträge in überarbeiteter Form in einem ZfK-Report erscheinen zu lassen.

M. Böttger

Eingegangen am 5. 9. 1983

Isotopenpraxis 20 (1984) 2, S. 75 bis 80

Buchbesprechungen

Management of Wastes from Uranium Mining and Milling Proceed. Internat. Symp., Albuquerque, 10—14 May 1982

Vienna: IAEA 1982. 735 S., div. Abb. u. Tab. (Proc. Series, STI/PUB/622); karton. OeS 1200,—

Für die weitere Entwicklung der Kernenergie-Wirtschaft ist die Behandlung der radioaktiven Abfälle, die im Brennstoffzyklus entstehen, von wesentlicher Bedeutung. Dabei sind nicht nur die Probleme der Behandlung von abgebrannten Brennstoffelementen zu sehen, die in der Öffentlichkeit besonderes Interesse finden, sondern auch die Fragen der richtigen Behandlung der Rückstände aus der bergmännischen Gewinnung von Uraniumerzen und ihrer Verarbeitung, die mit dem Blick auf sehr lange Zeiträume beantwortet werden müssen.

Nachdem die IAEA und die NEA schon mehrere Treffen und Veröffentlichungen zu diesen Fragen veranlaßt hatten, wurde 1982 ein Symposium veranstaltet, das sich ausschließlich mit diesem Gegenstand befaßte. Es nahmen 191 Fachleute aus 16 Staaten und von 2 internationalen Organisationen teil. Die Vorträge wurden in 7 Sektionen gegliedert: Übersichten über nationale Programme (4 V.; USA, Kanada, Australien, Frankreich); langfristige Programme und Kriterien (4 V.); Bewertung von Programmvorschlägen (5 V.); Beschreibung der Eigenschaften der Rückstände (5 V.); Behandlung der Abwässer und der Verarbeitungsrückstände (4 V.); Stillegung von Anlagen und Rekultivierung (7 V.); Bewertung der radiologischen Wirkungen (3 V.); Überwachung der Umgebung (5 V.); Strahlungsmessungen (4 V.); nationale Forschungs- und Entwicklungsprogramme (4 V.). Die Tagung wurde mit einer Ansprache von *F. E. Coffman* (USA) eröffnet und mit Schlußbemerkungen von *R. A. Scarano* (USA) beendet. Eine Rundtisch-Diskussion diente der Klärung des erreichten Standes und der künftigen Aufgaben.

Auf einige bemerkenswerte Ergebnisse soll hingewiesen werden: Die Rückstände des Bergbaus und der Aufbereitung enthalten auch nicht-radioaktive Stoffe, die eine Belastung der Umwelt darstellen. Hierauf ist ebenso Rücksicht zu nehmen wie auf die radioaktiven Komponenten. Staatliche Regelungen sind in den letzten Jahren entsprechend dem neuen Wissensstand ergänzt oder überarbeitet worden. Geomorphologische Untersuchungen sind erforderlich, um den langfristigen Einfluß der exogenen und endogenen Kräfte auf Halden und Absetzbecken einzuschätzen. Besondere Aufmerksamkeit gilt der Verhinderung der Kontamination der Oberflächen- und Grundwässer. Die Abscheidung von Radium-226 und Polonium-210 erleichtert die Behandlung der Rückstände. Aufbereitungsrückstände können stabilisiert werden. Methoden der Umgebungsüberwachung. Beobachtungen über das Verhalten von Radon und Radium.

Die umfangreiche Veröffentlichung läßt deutlich werden, daß die internationale wissenschaftliche Öffentlichkeit ein klares Verständnis für die möglichen Gefährdungen, die von den Rückständen des Bergbaus und der Aufbereitung ausgehen können, und dadurch Mittel für die wirksame Beherrschung der Probleme gewonnen hat, wie es *Scarano* im Schlußwort ausdrückt.

Der angesprochene Leserkreis wird die Vorträge mit Interesse aufnehmen. *K. H. Köhler*

Identifiability of State Space Models With Applications to Transformation Systems

E. Walter. Berlin-Heidelberg-New York: Springer-Verlag 1982. VIII, 202 S., 56 Abb., 1 Tab., 162 Lit. (Lecture Notes in Biomathematics, 46); karton. DM 32,50; US $ 13.60

Hinter dem trockenen Titel verbirgt sich ein sehr aktuelles und teilweise recht kontrovers diskutiertes Problem. Es sei die Aufgabe gestellt, ein mathematisches Modell eines realen Vorganges zu erarbeiten welches nicht nur das beobachtete Input-Output-Verhalten widerspiegeln, sondern auch ein besseres Verständnis für die inneren Ursachen des Vorganges ermöglichen soll. Bei der Durchführung dieses Vorhabens ist der Test auf Identifizierbarkeit des vorgegebenen Modells ein wesentlicher Anfangsschritt. In der Tat: Ist die Struktur, die für das Modell gewählt wird, auf der Basis geplanter Messungen nicht identifizierbar, dann repräsentieren, ungeachtet der möglichen Qualität der Experimente, verschiedene Modelle mit gleicher Struktur, aber unterschiedlichen Parameterwerten die beobachtete Realität gleich gut. Es liegt auf der Hand, daß dann Parameterwerte, die durch irgendein Rechenverfahren ermittelt werden, höchst fragwürdig, wenn nicht bedeutungslos sind — ein kritisches Problem für jede Nutzung der Informationen aus solchen strukturellen Modellen.

Die Monographie baut auf dem Formalismus der Transformationssysteme auf, eine von *P. Delattre* eingeführte Erweiterung der Compartmentmodellierung (die als eine lineare Theorie aufgefaßt wird), welche von Bedeutung für die verschiedensten Probleme auf dem Gebiet der Biologie, Chemie, Pharmakodynamik, Ökologie, Ökonomie, Populationsdynamik oder Regelungstechnik ist. Nach einer Einführung in diesen Formalismus wird das Problem der strukturellen Identifizierbarkeit von verschiedenen Gesichtspunkten aus diskutiert und einer Lösung zugeführt (Erörterung von Methoden zum Test auf lokale und globale Identifizierbarkeit). Durch das gesamte Buch hindurch werden immer wieder nichtlineare Modelle in die Betrachtung einbezogen, weit mehr, als dies in vergleichbaren Publikationen (die bis 1981 berücksichtigt wurden) bislang der Fall gewesen ist. So ausführlich und prägnant ist auch noch kaum auf die Bedeutung von Tracerexperimenten als Basis für die Modellierung nichtlinearer Prozesse hingewiesen worden. Beispiele für die Anwendung der dargestellten Methodik stammen aus dem Gebiet der Stoffwechseluntersuchung mittels Tracern (Iodumsatz in der Ratte; ^{125}I), aber auch aus der Pharmakodynamik und der chemischen Kinetik.

Die Lektüre des Buches verlangt nicht unerhebliche mathematische Vorkenntnisse. Es wird also vor allem dort von Nutzen sein, wo in interdisziplinärer Zusammenarbeit Mathematiker, Physiker, Physikochemiker gemeinsam mit Biologen oder Medizinern an die Erarbeitung mathematischer Modelle realer, durch Messungen erfaßter Vorgänge gehen. Für dieses Vorhaben liefert der Band eine Fülle wertvoller Informationen und Erkenntnisse.

E. Winkler

Angewandte Zeitreihenanalyse

R. K. Otnes, L. Enochson. Aus dem Engl. von W. I. Chochlova. Moskau: Isdatelsvo „Mir" 1982. 428 S., zahlr. Abb. (in Russisch); Rubel 2,— ; M 10,—

In den verschiedensten Bereichen von Wissenschaft und Technik finden immer stärker Methoden der Zeitreihenanalyse Anwendung. Der vorliegende erste Band einer Reihe stellt grundlegende Methoden zur Behandlung derartiger Probleme dar.

Während im 1. Kapitel eine Reihe wichtiger Begriffsdefinitionen eingeführt werden, sind im 2. Kapitel die notwendigen Voraussetzungen seitens der Wahrscheinlichkeitstheorie und Statistik zusammengestellt. Das 3. Kapitel ist der Datenbereitstellung und -vorbehandlung gewidmet. Dabei geht es z.B. um die Gewinnung der Daten aus analogen und digitalen Signalen, die Aufbereitung für eine Darstellung in einer EDVA und Trendelimination. Im 4. Kapitel wird die Konstruktion unterschiedlicher digitaler Filter vorgestellt, für einige davon wird im 5. Kapitel eine rechentechnische Realisierung angegeben. Außerdem werden weitere praktische Aspekte der digitalen Filterung behandelt. Die zentrale Stellung der *Fourier*transformation wird im 6. Kapitel herausgearbeitet. Es werden sowohl der theoretische Hintergrund als auch rechentechnische Realisierungen dargestellt. In den folgenden Kapiteln werden für das Gebiet der Zeitreihenanalyse bedeutsame Funktionen beschrieben.

Kapitel 7: Kovarianz und Faltung, Kapitel 8: Kreuzspektraldichte und Leistungsspektrum, Kapitel 9: Übertragungsfunktion und Kohärenzfunktion.

Die Autoren wählen als Ausgangspunkt ihrer Darlegungen 2 Disziplinen, die eng miteinander verknüpft, aber nicht identisch sind — die Zeitreihenanalyse und die Signalverarbeitung. Beiden gemeinsam ist die Nutzung der hier vorgestellten Methoden.

Im Buch ist das Arbeitsgebiet der Autoren, der Standpunkt, von dem aus sie die Methoden darstellen, nämlich von der Seite der Signalverarbeitung, der Elektrotechnik und Akustik, klar erkennbar. Dort liegt auch zumeist die Quelle ihrer Daten. So ist auch die Auswahl der vorgestellten Methoden zu verstehen, die eine Wechselbeziehung zur Meß-, Steuer- und Regelungstechnik besitzen, wie z.B. die Behandlung digitaler Filter und der Übertragungsfunktion. Methoden der Zeitreihenanalyse, die stärker von der mathematischen Statistik ausgehen, wie z.B. Methoden zur Analyse im Zeitbereich, AR-, MA-, ARMA- u. ä. Modelle oder spezielle Spektralschätzungen, sind hier nicht oder nur kurz vorgestellt.

Für eine Einführung in die Problematik der Zeitreihenanalyse, insbesondere auf der Basis von Daten, die aus anderen Bereichen als den hier behandelten stammen, ist das Buch nicht gedacht und auch nicht geeignet.

Nützlich ist das Buch für jeden, der mit dem Anliegen der Zeitreihenanalyse vertraut, eine fundierte Darstellung der hier enthaltenen Methoden sucht. Dabei wird es für Leser aus dem Ingenieurbereich etwas leichter lesbar sein als für den statistisch orientierten. *M. Remer*

Aufbau und Anwendung dialogfähiger informationslogischer Systeme

M. G. Pkhhorelishvili. Tbilisi: Mecniereba 1982. 62 S., 1 Abb., 37 Lit. (in Russisch); Broschur

Die Monographie des Mitarbeiters des Rechenzentrums der Akademie der Wissenschaften der Grusinischen SSR, *M. G. Pchorelishvili*, behandelt einen sehr wichtigen Aspekt der modernen Informationsverarbeitung, die Programmentwicklung der informationslogischen Systeme. Diese Informationssysteme stellen zur Zeit die höchste Entwicklungsstufe der sog. faktographischen Systeme dar. Dank einer Reihe von Funktionen, wie z.B. die Bearbeitung von Fragmenten der natürlichen Sprache, Sammlung, Suche und Modifizierung der Datenbestände über Gegenstände, Erscheinungen und Prozesse, logische Analyse der Modelle der realen Prozesse und Lösungen höherer logischer Aufgaben, stellen sie echte dialogfähige Auskunftssysteme dar, bei welchen die EDVA in die Lage versetzt wird, neue Informationen aus den gespeicherten Daten zu gewinnen. Der Autor beschreibt den Aufbau und die Funktionen zweier derartiger Systeme, DILOS und TBILOS, welche bereits in einigen wissenschaftlichen Einrichtungen der UdSSR zur Lösung von informationslogischen Aufgaben eingesetzt wurden. Beide Systeme sind in der Programmiersprache LISP geschrieben und auf BESM6 Anlage sowie auf einer EDVA des ESER realisiert. DILOS wird beschrieben als ein dialogorientiertes informationslogisches System zur intellektuellen Vermittlung zwischen dem Endnutzer und den traditionellen Mitteln des Rechners bei der Lösung von Aufgaben. Das System besteht aus einer Modelldatei, wo Daten über Gegenstände oder Prozesse in frame-Struktur gespeichert werden, einem linguistischen Prozessor, in welchem die Sätze der natürlichen Sprache in eine Sprache des formellen Interface umgewandelt werden sowie aus drei Arbeitsprozessoren (Informationsprozessor, logischer Prozessor, Berechnungsprozessor), in welchen die Informationen verarbeitet und synthetisiert werden.

Das System TBILOS stellt eine Weiterentwicklung des DILOS dar. Es besteht aus folgenden Prozessoren: linguistischer frame-Prozessor (LINGFRES), Modifikator, Funktion FIND, logischer Prozessor. Mit Hilfe beider Systeme wurde eine Reihe von Funktionen getestet, welche zur Entwicklung komplizierter informationslogischer Systeme dienen. Dazu gehören z.B.: das Verstehen der natürlichen Sprache, Formen der Wissensdarstellung, Lösung von schwachstrukturierten, intellektuellen Aufgaben, Automatisierung von Prozessen der Entscheidungsfindung. Eine Reihe von Beispielen und Programmschritten demonstriert sehr instruktiv die Arbeitsweise der Systeme und die in ihnen liegenden Möglichkeiten. Das Buch kann für die Entwickler der faktographischen und informationslogischen Systeme sowie für die auf die Datenbankproblematik orientierten Programmierer von Interesse sein.

L. Jankowski

Weltreich der Physik
Von Galilei bis Heisenberg

A. Hermann. Esslingen am Neckar: Bechtle Verlag 1980, 392 S., 30 Abb. im Text, 27 Abb. auf Kunstdruck; Leinen DM 34,–

Zu den vielen bisher im Laufe der Jahre schon erschienenen Büchern über die Naturwissenschaften und deren Entwicklung gesellt sich mit dem „Weltreich der Physik" ein weiteres. Es handelt sich dabei aber weder um eine bloße Wiederholung von auf einem speziellen Gebiet bereits Vorhandenem, nur mit anderen Überschriften und Worten, noch um eine Art Ergänzung oder Fortsetzung einschlägiger Werke.

Der Autor umreißt in zehn Kapiteln an Hand der bedeutsamsten Entdeckungen von 30 der namhaftesten Wissenschaftler (von *Galilei* über *Helmholtz*, *Faraday*, *Hertz*, *Rutherford* und *Einstein* bis *Heisenberg*) die Entwicklung der Physik. Er behandelt den umfangreichen und je nach Fachkenntnis des Lesers zum Teil nicht unbedingt leicht überschaubaren Stoff interessant erzählend und sehr verständlich beschreibend. Es werden neben den wesentlichsten wissenschaftlichen Leistungen der großen Physiker (unter Verwendung unterschiedlichster Belege wie Briefe, Tagebücher, Notizen über Vorträge und Diskussionen u. a.) deren Motivationen, ihr Werdegang, die zum Teil beträchtlichen Probleme, mit denen sie sich auseinanderzusetzen hatten, Rückschläge und Erfolge und nicht zuletzt ihr Auftreten in der Gesellschaft umrissen. All das ist vom Verfasser, sehr gelungen miteinander verknüpft, zu einem unterhaltsamen und in vielfältiger Weise belehrenden Buch zusammengestellt worden. Es ist mit zahlreichen historischen Photographien, Bildern von fundamentalen Versuchsanordnungen und Abdrucken von für verschiedene Entdeckungen entscheidenden Labortagebuch- oder anderen Aufzeichnungen, Titelseiten oder Teilen wichtiger Arbeiten und Publikationen illustriert. *J. Holzhey*

Atomspektrometrische Spurenanalytik
Vorträge beim Kolloquium „Spurenanalytik mit Atomspektroskopie" in Konstanz, 7. bis 9. April 1981

Hrsg.: B. Welz. Weinheim – Deerfield Beach – Florida – Basel: Verlag Chemie 1982. XVI, 564 S., 295 Abb., 137 Tab., 456 Lit.; brosch. DM 148,–

Die atomspektrometrische Spurenanalytik, besonders die Atomabsorptionsspektrometrie ist in den letzten zehn Jahren für die Elementspurenanalytik im ppm- und sub-ppm-Bereich zum Routineverfahren der Wahl geworden. Das gilt besonders dann, wenn es darum geht, umfangreiche Serienanalytik in ökonomisch vertretbarem Rahmen durchzuführen, wie sie bei der Bewältigung analytischer Probleme in den Bereichen der Umweltforschung, Biologie, Medizin, Hydrologie und Geologie zur statistischen Absicherung der gewonnenen Informationen notwendige Voraussetzung ist.

Der vorliegende Band enthält 55 Beiträge eines in Konstanz veranstalteten Kolloquiums zur Spurenanalytik mit Atomspektrometrie. Schwerpunkte dieses Kolloquiums waren neben dem vorgestellten methodisch-apparatetechnischen Fortschritt (Graphitrohrofen-, Hydrid- und Kaltdampftechnik bei der AAS; Inductively Coupled Plasma-Technik (ICP) bei der AES) und den Problemen der Probenvorbereitung und -anreicherung vor allem die Analytik biologischer Materialien einschließlich der klinischen Analytik (Problemelemente: Cu, Zn, Se, Cd, Hg, Tl, Pb; Objekte: Urin, Blut, Gewebe, Knochen); die Umweltanalytik (z. B. ökotoxische Bewertung von Stäuben; Schwermetallanalytik in Luft, Wasser, Abwasser, Klärschlamm, Müllkomposten); die Analytik in der Metallurgie, Geologie und Erdölklassifizierung (z.B. Spurenelementbestimmung in Rohölen als geochemische Methode; Gesteins- und Mineralienanalyse durch AAS; Einsatz der AAS zur geochemischen Kartierung); die Analytik von Lebensmitteln und pharmazeutischen Produkten (z.B. Se in Lebensmitteln; Spurenelemente in Gemüse).

In zwei Beiträgen wird die Atomspektrometrie benutzt, um Reinheitsbestimmungen von U-, Pu-, Am-, Cm- und Pa-Proben durchzuführen (z.B. Wiederaufbereitungsanlage Karlsruhe; Einsatz eines AAS zur Bestimmung von >30 Spurenelementen in U- und Pu-Lösungen; Zeitbedarf 24 h).

Die vorgestellten Beiträge stellen den derzeitigen Weltstand auf diesem Gebiet dar (zit. Literaturstellen: 456; davon <1960: 18; 1960–1970: 58; 1971–1980: 330; $>$ 1980: 50).

Beeindruckend ist der erreichte Stand der Gerätetechnik. Deutlich erkennbar ist der Trend zur Multielementanalytik (Verfahren der Vorwärtsstreuung, Mehrlementhohlkatodenlampen).

Für die Einarbeitung in das Arbeitsgebiet der Atomspektrometrie kann dieser Tagungsband fast als Kompendium empfohlen werden. Hier erhält der Leser nicht nur in beeindruckender Weise Informationen zur Atomspektrometrie und zur experimentell-technischen Durchführung, sondern auch viele nützliche Hinweise zu dem auch für andere analytische Verfahren wichtigen Problemkomplex „Probenvorbereitung und Spurenelementanreicherung".

J. Flachowsky

Atomvorstellungen im 19. Jahrhundert

Ch. Schönbeck (Hrsg.). Paderborn – München – Wien – Zürich: Ferdinand Schöningh 1982. 96 S., 43 Abb., einige Tab.; karton. DM 13,80

Diese kleine Schrift entstand aus Vorträgen, die anläßlich der Frühjahrstagung der Deutschen Physikalischen Gesellschaft im März 1980 zu diesem Thema gehalten wurden, ergänzt durch einen Beitrag von der Herausgeberin der Schrift, *Ch. Schönbeck.*

Es sind folgende Beiträge enthalten: *Claus Priesner,* Zur Entwicklung der Atom- und Moleküldefinition in der Chemie im 19. Jh.; *Karl von Meyenn,* Engpässe in der Atomtheorie des frühen 19. Jh.; *M. Norton Wise,* Atomism and Wilhelm Weber's concept of force; *Charlotte Schönbeck,* Atommodelle um 1900.

Es ist erstaunlich, wie mühevoll und von Irrwegen begleitet sich der Begriff des Atoms und die Atomtheorie entwickelt haben, wobei das 19. Jh. eine entscheidende Etappe bildete. Was heute jeder Schüler einer höheren Lehranstalt als fast selbstverständlich anzusehen geneigt ist, hat viele Diskussionen und Meinungsstreitigkeiten durchlebt, wovon diese kleine Schrift berichtet und einen kleinen Eindruck davon wiedergibt. Allein das sorgsame Auseinanderhalten der Begriffe Atom und Molekül hat eine Geschichte hinter sich. Sehr bemerkenswert (und wohl kaum in einer anderen Darstellung so umfassend zusammengestellt) ist die Geschichte des Atommodells um 1900 abgehandelt. Man lernt im allgemeinen nur einige wenige Ansichten von *Thomson,* aber nicht, daß es allein 4 unterschiedliche Modelle von *Thomson* gab.

Zwei Bemerkungen scheinen notwendig zu sein. Es ist immer wieder zu finden, daß der Schweizer Lehrer *J. J. Balmer* so knapp abgehandelt wird. Er hat 1885 7 Seiten in den Annalen der Physik veröffentlicht, und seither ist sein Name mit der bekannten Serienformel des Wasserstoff-Linienspektrums verbunden. Wer aber war *J. J. Balmer,* wie kommt ein Lehrer auf eine solche Formel, was hat er eigentlich unterrichtet? Solche und ähnliche Fragen tauchen beim Lesen dieses Namens auf; und sein Einfluß auch auf die Herausbildung des Atommodells dürfte nicht unwesentlich gewesen sein. Die andere Bemerkung betrifft die Arbeiten und Äußerungen von *William Crookes.* Natürlich ist sein Name mit der bekannten *Crookes*schen Mühle in den Schaufenstern von Optikern verbunden, aber außerdem hat er Gedanken geäußert, die seiner Zeit weit vorauseilten. So hat er 1886 in Birmingham u. a. gesagt: „Ich stelle mir daher vor, daß, wenn wir sagen, das Atomgewicht des Ca z.B. sei 40, wir in Wirklichkeit die Tatsache ausdrücken, daß, während die Mehrzahl der Ca-Atome tatsächlich das Atomgewicht 40 haben, es doch eine nicht geringe Zahl gibt, die durch 39 und 41 repräsentiert werden, eine geringere Zahl durch 38 oder 42 usw.". Erkennen wir nicht heute darin die isotope Zusammensetzung der Elemente wieder, die erst 1911 in der jetzigen Bedeutung ausgesprochen wurde? *William Crookes* hat aber auch gesagt: „Wir haben gesehen, daß in einigen ihrer Eigenschaften die strahlende Materie ebenso materiell ist wie dieser Tisch, während sie in anderen Eigenschaften fast den Charakter strahlender Energie annimmt." Sind dies nicht die Materiewellen von *de Broglie* und die Wellenmechanik von *Schrödinger? Crookes* hat sich auch mit Okkultismus beschäftigt; vielleicht hat er dadurch seinen Zeitgenossen gegenüber an Glaubwürdigkeit eingebüßt, aber bei *Lenard* hat er einen sehr nachhaltigen Eindruck hinterlassen.

Es ist immer lehrreich und anregend, eine Entwicklungsgeschichte nachzuvollziehen und die Herausbildung eines Wissenszweiges zu erleben, vor allem, wenn der Leser weiß, was zum Schluß herauskommen soll. So verstanden ist diese kleine Schrift wertvoll und für unterschiedliche Leserkreise sehr geeignet und empfehlenswert. *H. Hübner*

Physikalische Chemie
Teil II: Aufbau und Eigenschaften der Kerne, Atome und Moleküle

G. M. Barrow. Übers. aus dem Engl. u. Bearb. G. W. Herzog. 3. Aufl. (Nachdruck). Wien: Bohmann-Verlag u. Braunschweig/Wiesbaden: F. Vieweg & Sohn 1980. 295 S., 114 Abb., 21 Tab., 25 Lit. (Reihe Uni-Text); brosch. DM 26,80

Der zweite Teil dieses dreiteiligen Lehrbuches der Physikalischen Chemie enthält die Kapitel 10–15. Er behandelt den Aufbau der atomaren und molekularen Bausteine der Materie und leitet daraus einige ihrer makroskopischen Erscheinungsformen ab. Basierend auf einer Einführung in die Grundlagen der Quantenmechanik im Teil I werden zunächst am Beispiel des Wasserstoffatoms die wichtigsten Aussagen der Quantentheorie des Atomaufbaus abgeleitet. Darauf aufbauend folgt eine quantenmechanische Erklärung des Wesens der chemischen Bindung und der Zusammenhänge zwischen Bindung und Molekülstruktur. Der ihrer Bedeutung angemessene Raum wird in drei Kapiteln den Methoden der Strukturuntersuchung eingeräumt. So wird aus den Grundlagen der Quantenmechanik abgeleitet, welche Strukturaussagen die Molekülspektren aller Wellenlängenbereiche zulassen. Wie diese ergänzt werden können, wird durch die Behandlung der Grundlagen und Möglichkeiten der Beugung von Elektronen- und Röntgenstrahlen an Gasen und kondensierten Stoffen erläutert. Die Methoden der Strukturuntersuchung werden abgeschlossen durch die Behandlung der elektrischen und magnetischen Eigenschaften der Moleküle, die sich aus der Orientierung permanenter molekularer Dipole und der Polarisierbarkeit von Molekülen in äußeren elektromagnetischen Feldern ergeben. Sehr knapp werden der Aufbau der Atomkerne und die elementarsten Grundlagen der Radioaktivität dargestellt. Um die didaktische Entwicklung der zahlreichen theoretischen Ableitungen überschaubar zu halten, werden in bewährter Weise für das Verständnis notwendige spezielle mathematische Grundlagen im Anhang untergebracht.

Das Buch wendet sich an Studierende der Chemie, Verfahrenstechnik und Physik und setzt Grundkenntnisse in diesen Fächern und in höherer Mathematik voraus. Es ist als einführendes Lehrbuch zum Gebrauch neben Vorlesungen geeignet und enthält zu jedem Kapitel einige Kontrollfragen und Aufgaben. Der didaktische Aufbau und die ausgewählten Anwendungsbeispiele sind eine gute Empfehlung auch für diesen Band.

G. Müller

Physikalische Chemie
Teil III: Mischphasenthermodynamik, Elektrochemie, Reaktionskinetik

G. M. Barrow. Übers. aus dem Engl. u. Bearb. G. W. Herzog. 3. neubearb. Aufl., Wien: Bohmann-Verlag u. Braunschweig/Wiesbaden: Vieweg & Sohn 1979. 378 S., 147 Abb., 63 Tab. (Reihe Uni-Text); kart. DM 34,–

Dieser 3. Band des Lehrbuches für physikalische Chemie behandelt in 9 Kapiteln die Themen Festkörper, Flüssigkeiten, Zustandsdiagramme ein- und mehrkomponentiger Systeme, Mischphasenthermodynamik, Elektrochemie, Makromoleküle und die homogene und heterogene Kinetik, einschließlich der Adsorption und Katalyse. Wie es einem solchen Lehrbuch für Chemiker, Verfahrenstechniker und evtl. Physiker ab dem 3. Semester zukommt, wird sehr viel Wert auf ein grundsätzliches Verständnis – man ist geneigt zu sagen auf ein Einfühlungsvermögen – des zu behandelnden Stoffgebietes gelegt, wobei flankierend und verstärkend die Ableitungen der notwendigen Gleichungen sehr ausführlich dargeboten werden, die Abbildungen gut überschaubar sind und die ausgewählten Beispiele auch im Sinne dieses Anliegens wirken.

Einige kleine Wünsche bleiben offen: Im Vorwort zu dieser 3. Auflage heißt es, daß die SI-Einheiten eingeführt wurden; dies trifft nicht für die Rechenbeispiele zu, die noch größtenteils mit Torr, Atm. u. a. Einheiten abgehandelt werden. Die Kettenreaktionen werden an zwei Stellen berührt – S. 281 u. S. 298 –, aber es wird nicht das *Bodenstein*sche Stationaritätsprinzip erwähnt, wie auch auf S. 302 nicht von der *Michaelis-Menton*-Gleichung und der *Michaelis*-Konstanten $K_M = \text{const.}$ gesprochen wird. Die Diffusion und vor allem der mathematische Apparat der Behandlung von Diffusionsproblemen (Randwertprobleme) sind etwas knapp und nur phänomenologisch abgehandelt. Die Selbstdiffusion ist nicht nur ein ungerichteter Diffusionsvorgang, sondern kann in isotopen Systemen direkt beobachtet werden (H_2O/D_2O; $H_2O/H_2{}^{18}O$; H_2O; T_2O oder Spinechomethode). Die Behandlung isotoper Systeme erscheint insgesamt zu knapp, beispielsweise sind Gemische unterschiedlich isotoper Spezies ideale Zweikomponentengemische, die Kinetik des Isotopenaustausches folgt unter Vernachlässigung von Isotopieeffekten einer Reaktion 1. Ordnung. Bei der Formel auf S. 295 zur Austauschreaktion des asym. Phenol-sec. butylketons sind H^+ bzw. D^+ wirksam.

Das didaktisch ausgewogene Lehrbuch dürfte auch künftig einen breiten Leserkreis finden, wenn auch vielleicht einige neuere Arbeitsrichtungen (nicht-isotherme Reaktionsverläufe, Modellbetrachtungen einfacher Flüssigkeiten, Lösungsverhalten überkritischer Gase oder chemischer Bindungen an Oberflächenfestkörpern beispielsweise) zu Lasten anderer Teilgebiete, die kürzer beschrieben werden können, Erwähnung und Aufnahme finden sollten.

H. Hübner

Mathematisch-statistische Methoden in der praktischen Anwendung

E. Renner. 2. Auflage. Berlin und Hamburg: Verlag Paul Parey 1981. 112 S., 11 Abb., 62 Tab. (Pareys Studientexte); karton. DM 24,–

Der Autor wendet sich in seinem Vorwort vor allem an Nichtmathematiker, denen er den Zugang zur Nutzung mathematisch-statistischer Methoden erleichtern will. Er verzichtet dabei bewußt auf mathematische Ableitungen und gibt, ausgehend von den Berechnungsformeln, die numerischen Rechengänge und Zwischenergebnisse ausführlich anhand von Anwendungsbeispielen wieder. Nach einer Einleitung werden folgende Themen behandelt: 2. Kap.: Häufigkeitsverteilungen, 3.: Vergleich von Verteilungen, 4.: Mittelwert, 5.: Varianzanalyse, 6.: Korrelationsrechnung, 7.: Meßfehler von Bestimmungsmethoden, 8.: Ausreißernachweis, 9.: Anwendung von Taschenrechnern. Der Autor erhebt den Anspruch, „trotz der bewußten Vereinfachung in der Darstellung der mathematisch-statistischen Methoden alle Methoden zu behandeln, die ein allgemeines Interesse bei der Auswertung von Untersuchungsdaten beanspruchen können". Dieser Vorsatz scheint dem Rezensenten unter Beachtung des vorgegebenen Umfangs kaum realisierbar, wobei aber in weiteren Auflagen zumindest solch elementare Methoden wie z.B. Korrelations-/Regressionsanalyse für mehr als eine Einflußgröße oder auch Hinweise auf weitere Verteilungstypen Berücksichtigung finden müßten. Die im 9. Kapitel getroffene Aussage, daß „die aufgeführten Formeln und Verfahren statistischer Methoden zum selbständigen Programmieren in vollem Umfang geeignet sind", ist sicherlich richtig, sollte aber nicht als unbedingte Aufforderung zum Programmieren betrachtet werden. Vielmehr sollte man, nachdem man sich mit dem Anliegen der statistischen Methode vertraut gemacht hat — hierin liegt auch der eigentliche Wert der ausführlichen Darlegungen der Rechengänge —, auf getestete Standardsoftware zurückgreifen.

Das Büchlein ist für Statistik-Laien als elementare Einführung in einige Teilbereiche geeignet, indem es die Methoden als „Kochrezepte" darstellt. Es wird dem Anspruch gerecht, ein erstes Verständnis für die Anwendung statistischer Methoden zu wecken, wobei aber ein kurzer Ausblick auf Weitergehendes vermißt wird. Als für den „fortgeschrittenen Statistiker zum schnellen Nachschlagen" geeignet, scheint es überfordert.

M. Remer

Trace Element Analytical Chemistry in Medicine and Biology, Volume 2
Proceedings of the Second International Workshop, Neuherberg, Federal Republik of Germany, April 1982

Hrsg. P. Brätter u. P. Schramel. Berlin — New York: de Gruyter 1983. 1189 S., 264 Abb., 240 Tab., 1565 Lit.: geb. *DM 280,—*

Schon zwei Jahre nach Durchführung des ersten internationalen Workshops über die Spurenelementanalytik in der Medizin und Biologie (im April 1980 in Neuherberg, BRD) hat die zweite Arbeitstagung dieser Art (im April 1982 am gleichen Ort) stattgefunden, und so, wie aus der ersten Tagung der erste Band der Serie „Trace Element Analytical Chemistry in Medicine and Biology" hervorgegangen ist, haben die Vorträge und Diskussionen der zweiten Tagung ihren Niederschlag in diesem Band gefunden. Er ist noch umfangreicher und vielseitiger als der erste. Hatten im ersten Band überwiegend methodische Fragen der Spurenanalytik im Vordergrund gestanden, so enthält der vorliegende neben der Analytik, die keineswegs zu kurz kommt, in erfreulich hohem Maße Beispiele für die erfolgreiche Anwendung moderner Analysenverfahren mit den entsprechenden Auswertungen und Schlußfolgerungen. Auf der zweiten Arbeitstagung im April 1982 sind in 5 Sektionen insgesamt 92 Vorträge gehalten worden. Außer der Analytik, der stärksten Sektion, waren der Bedarf des Menschen an Mineralstoffen und Spurenelementen, die Biochemie der Spurenelemente, ihre chronische Wirkung sowie die Diagnose von Spurenelementmangelzuständen Grundthemen der Tagung.

Die hohe Beteiligung, die Fülle des dargebotenen Materials und die Vielseitigkeit der Themen sind ein Ausdruck für das weltweite Interesse, das die Mineralstoffe und noch mehr die Spurenelemente seit zwei oder drei Jahrzehnten beanspruchen. Eingeleitet und ermöglicht worden ist diese Entwicklung durch die gewaltigen Fortschritte auf dem Gebiet der Spurenelementanalytik, aber auch neuere Methoden der Enzymatik und der Proteinchemie haben wesentlich zur Vertiefung der Kenntnisse der Mineralstoffe und Spurenelemente im biologischen Geschehen beigetragen. Auf keinem anderen Gebiet der Biochemie und der Ernährungslehre (sei es Pflanzen-, Tier- oder menschliche Ernährung) sind in letzter Zeit so viele neue Erkenntnisse hinzugekommen wie auf dem Gebiet der Mineralstoffe und Spurenelemente. Erinnert sei nur an den Nachweis der Essentialität vieler Spurenelemente, deren Vorkommen im menschlichen Organismus früher gar nicht bekannt war oder die als ausschließlich toxisch galten. Auf keinem anderen Gebiet sind aber auch noch so viele Fragen offen wie auf diesem. Man denke dabei nur an die Fragen des täglichen Bedarfs bzw. der höchstzulässigen täglichen Zufuhr, an Verfügbarkeit, Verteilung, gegenseitige Beeinflussung und Einflußnahme auf die verschiedensten Lebensprozesse. Die Bereitstellung an neuen Informationen jeglicher Art ist vielfältig, und ebenso von Vielfalt geprägt sind die Themen, die in diesem Werk behandelt werden.

Bei der Vielzahl der Themen und der zufälligen Auswahl der Verfasser (die Vorträge sind so abgedruckt worden, wie sie angemeldet und auf der Tagung vorgetragen worden sind), ist es ganz natürlich, daß die Sammlung ziemlich heterogen ist. Die Skala reicht vom Übersichtsreferat mit einem hohen Grad an Verallgemeinerung und Abstraktion bis zum Spezialvortrag mit eng begrenzter Thematik, aber detailierter Darstellung. Eines haben jedoch alle Artikel gemeinsam: Sie sind von hoher Aktualität und in dieser Aktualität von hohem Informationswert.

In irgendeiner Form kommen fast alle lebensnotwendigen oder überwiegend toxischen Elemente in diesem Werk vor, wobei es selbstverständlich Schwerpunkte gibt. Von den essentiellen Mineralstoffen werden immer noch Eisen und Zink vorrangig bearbeitet, von den toxischen Spurenelementen stehen vor allem Cadmium, Chrom und Quecksilber im Mittelpunkt des Interesses, aber auch sonst weniger beachtete Elemente wie Zinn, Aluminium und Gold sind vertreten. In der Analytik dominieren die anspruchsvolleren Verfahren wie die Neutronen-Aktivierungsanalyse, die Tracertechnik, die Atomabsorptionsspektrometrie und die Sekundärionen-Massenspektrometrie.

Das Werk bietet in seiner Vielfalt sowohl dem Fachmann, der sich über den neueren Stand der Forschung und die an anderen Orten bearbeiteten Probleme informieren will, als auch dem mit der Materie noch nicht so tief Vertrauten viele Anregungen, Ratschläge und Denkanstöße, aber auch Vergleichsmöglichkeiten zur eigenen Arbeit und zur Einordnung seines eigenen Wissensstandes. Es ist nicht nur den experimentell, sondern auch den theoretisch arbeitenden Analytikern, Biochemikern und Medizinern sehr zu empfehlen.

F. Kretzschmann

ISOTOPENPRAXIS

Volume 20 1984 Number 2

Contents

Inhalt

Isotopenpraxis is indexed in Current Contents/Physical, Chemical & Earth Sciences, Science Citation Index, and INIS Atomindex